Fifth Force Assembling ©2000

by John Uldrich

A novel about time – for our time . . . .

China Gate Press ™
Shanghai, China, PRC

First Edition 2000

Book & Jacket Design by:
Wu Rui Qi

Author may be reached via Internet
< juldrich@public.sta.net.cn >
Website: < www.synsat.com >

ISBN 962 – 450 – 765 – 1/D.40942

Published by:
China Gate Press
9 – 10, #19, 1020 Rong
Si Ping Lu,Shanghai 200092, China PRC

86 – 21 – 6502 – 2232 (Phone & Fax)

Distributed by:
China Gate Press

Printed in the People's Republic of China

Future works in the *Quinn Quartet*

Silverplate III and the Sword, is about the third atomic bomb dropped on Japan on August 14, 1945. Based on historical fact bent to fiction – it involves Cold War foes, present day Japanese *yakuza*, *samurai* history and the *burakumin* – outcasts of Japanese society.

Red Seoul. First published in South Korea under the title, *Tangun Tunnel* in 1994, it is an espionage thriller based on North Korea's effort to take the South through military conquest via an elaborate tunnel mechanism.

The fourth work, untitled, about Shanghai of World War II （1938 – 45） and the present, is in progress.

Other Works by the Author:

Belle Isle: Gateway to Andersonville. An anthology of stories, reports and monographs about the first major prisoner – of – war camp of the American Civil War.

Blue Ribbon Barbeque: a collection of award winning recipes and favorites of friends and associates. Many unique recipes!

Blue Ribbon Burgers: over one hundred recipes for making the All – American Classic – the hamburger – a gourmet's delight!

Cooking With Brew: a complete treatise on cooking with beer and dozens of recipes from around the world that call for beer as an ingredient. Emphasis on use of famous microbrews.

Classic Hot Wings: dozens of recipes for preparing what is now an international favorite – chicken wings! All of John's cookbooks contain information on using herbs and spices from around the world – a reflection on his global travels.

Exercises For Those Too Tired, Busy or Lazy. Unique exercise techniques and regimens for *anyone & all ages*! From Asian countries – they have proven effective over the centuries!

For further information go to: < www.synsat.com >

Dedication

To my good friends on Minnesota's famed Iron Range, especially those north of the Laurentian Divide . . . .

Individually and as a society, they made my three years "*on the Range*" as a "*Boomer*" * – an epoch I fondly recall with warm memories and cherish over great distances.

I am particularly indebted to Jimmie Holmes, Buhl, Minnesota, hunting partner and comrade, for introducing me to the true '*heart & soul*' of the Iron Range.

His insights regarding the 56 ethnic groups on the Range – Italian & Finn – Native American and the last in – Vietnamese – contributed greatly to this work.

It was Jimmie who first introduced me to the Tower Soudan Mine. Together we explored its' dark chambers and slanting stopes thinking of growing mushrooms.

Mushrooms flourish not – our friendship does.

John Uldrich
Shanghai, China

* "*Boomer*" – local idiom for outsiders who come to work on the Range but then leave.

Roses Be Red
Violets Ere' Blue
Find Fifth Force
Nobel For You

Graffiti, Men's Third Floor Lavatory, Second Stall On Right, Physics Building, University of Minnesota, Minneapolis, Minnesota. (The same doggerel can be found in the men's *loo*, Physics Department, Cambridge University, England . . . the handwriting is quite similar).

FIFTH FORCE ASSEMBLING   (c) 2000

A novel by John Uldrich

"ACCESS DENIED."

Fluorescent yellow of the monitors radiating into the darkened room gave the two men, faces drawn, a sick, waxy hue. The twin monitors matched—both carried the same icon—the same mocking message.

The younger man stood up and turned· on a small desk lamp, shifting his gaze to the larger screen suspended over the twin PC's. Early morning rays of a late-fall sun filtered into the two-story frame house. A pterodactyl, crafted of balsa and tied to a string, drifted in silent, tethered flight. The occupants of the private residence stared intently at the screen in stony silence.

On the monitor, a small group of men and women, banded together, appeared to be listening to a speaker out of camera range. There was no audio. Of various ages, half of them in casual clothes, others in winter jackets and parkas—all wore uniformly grim faces.

The younger man looked down at the man in the wheelchair and finally broke the long silence. "Timing was letter perfect, Walt. Wolf Pack's finished breakfast while Bochert's team is coming down in the cage. We're locked out—everybody else is locked in. It's unreal!" Ted Theodopolus, junior member of the two-man programming team, glanced back at the screen. "Would you have believed Mohacker and Vidall were part of this?"

A cadaverous Doctor Walter LaMont was silent, rubbing a stubble chin with bony fingers. A small green oxygen tank stood beside the over-stuffed leather chair that doubled as a breakfast table. When he finally spoke, his voice was raspy. "Mohacker? Never! Too easy going! Vidall, our main Boolean man? No, I wouldn't have pegged either one as a mole."

Theodopolus shifted the remote controlled surveillance camera from the man called Mohacker and the Israeli-built Uzi he held at waist level, to the smaller, almond-eyed Vidall carrying an identical machine gun at port arms. He turned again to face LaMont whose eyes were closed, fingers of his right hand pinching the bridge of his nose.

"What do we do? Whoever's down there, they've got their hands on the entire system. We can't block the main-frame from here."

1

Eyes opening, staring at his uneaten breakfast, LaMont's left hand kneaded a cloth napkin nervously. "My immediate concern Teddy, is more pressing." He glanced up at the younger man.

"Meaning?" Theodopolus shifted, eyes darting between screen and LaMont.

"My hunch is, that one Doctor Theodopolus is on their acquisition list. Assuming they know we both work for the Company—that makes you a top priority! Mohacker or Vidall probably told them about my, ah, limited mobility, so I'm not of any great concern. But they discover the camera and trace it here—we're in deep trouble. That we're both watching this on closed circuit tells me they haven't figured out that you're here—not at your apartment—or enroute back."

"Don't forget Walt, this is the second week of the Quark Conference at Cambridge. No one knows I came back from England early to . . . ." The younger man's voice faded to silence.

The geophysicist looked up from his chair, a smile on his face. "Forgot about the conference Teddy—didn't forget the reason you came back . . . ."

LaMont coughed harshly. Theodopolus bolted to his side, pounding him on his back with practiced strokes, then placed a box of tissues at his side.

"Any suggestions?" he asked quietly.

LaMont looked up "Yes! Get the hell out of here! Our lines might be tapped, transceiver system compromised. Find the first pay phone and reach up the chain. As high as you go–Chen himself if you can. Tell them the Orchard's been taken over by parties unknown—that we need outside help. I . . . ." Voice trailing off, LaMont's head dropped from exertion.

Theodopolus squeezed LaMont's hand." Walt, you're doing just fine. Just fine."

Beads of perspiration dotted the older man's brow. He looked up. "Tell Washington how close we were to getting the final numbers. Your LMRP was perfect for tomorrow's run—and now the fruits of ten years work are . . . ." LaMont's words ended in a sigh.

*　　　*　　　*

Theodopolus left the rented, two and a half story house ringed by mature pines with a glance westward. He saw the wispy mare's tails

2

veil of clouds and remembered his grandfather, a Greek seaman's and the weather rhyme: "*The lower they get, the nearer the wet.*"

He looked northward, toward the massive Head-Frame on the ridgeline three-quarters of a mile away. Skeletal, angular, and ugly, Walt's description came to mind: "*Eiffel Tower without the Eiffel.*" It was the entrance-shaft into the Tower-Soudan Iron Mine. Millions of tons of high-grade ore remained locked between layers of granite, graywacke and Ely greenstone, he'd been informed. Cheaper, offshore ore had made continued mining at the extreme end of Minnesota's Iron Range too expensive. Now, during the summer months, it was a tourist attraction—and the location of a University of Minnesota research project—operating in tandem with England's Cambridge University. Theodopolus recalled the Greek name for the tumultuous era in which the rich ore was created. *Algos*—pain, *oros*—mountain, *geny*—origin, *mountain of pain.* Jutting into the sky like a huge, huge phallic symbol, the head-frame, with its massive sheave wheels, raised and lowered men and equipment into the mine. He shuddered—each descent was a complex moment of pleasure and pain—pleasure in that the rapid downward descent triggered sexual arousal—confinement in the iron cage—intense claustrophobia. Teddy shared the former thought with LaMont—kept the latter confined in his heart.

Five minutes later, in his rusting Fiat Spyder, vision of the Head Frame fading; he turned on to the main street of Tower, the village just west of Soudan. Inhabitants of the town were filing into the small cafes for their morning coffee break and freshly baked Finnish breakfast rolls and bread.

Two hours and three changes of pay phones to "*get up the chain,*" he headed west again, on a country road leading to the sod strip on the shores of Lake Vermillion. Chen, overall head of the project, was not at the Defense Analysis Institute. Now Theodopolus was to wait for a Company representative to personally debrief him. Two Army jeeps sat on the shoulder at the junction of Highway 169. The second in line only momentarily caught his attention because of its many antennas. Pre-occupied with thoughts of what was happening in the underground laboratory and LaMont's condition, he let the presence of the military vehicles slip from his mind.

3

As he disappeared beyond the first bend, one of the vehicles U-turned behind him.

Blue skies still washed over the southern edge of the Boundary Water Canoe Area but the darkening mass to the west seemed to be growing in size and intensity.

Minutes later, he was ensconced in the vintage silver Luscombe. The two-man aircraft had been his single luxury after accepting the research position.

Westerly winds blew over the dark waters of the lake. A sleek raven flitted through pine tops around the grass strip. The plane shuddered as a strong gust stressed the wings, putting pressure against rope tie-downs. Gloved hands resting on the yoke, he let his mind trace the long, low altitude flight taken three years earlier to the tip of the Florida Keys. A healthy, vibrant LaMont, his passenger in the side-by-side aircraft. The memory faded into sadness.

At three o'clock, cold and cramped, the thirty-year-old Theodopolus rationed out the third cup of thick Turkish coffee. Condensation formed on the window to his left. Putting the cup on the instrument panel, a translucent fan formed on the sloping windscreen. Outside vision fading, the young scientist closed his eyes and let his thoughts go back one week in time. In a small pub called the Slug & Lettuce, near London's Victoria Station, he remembered the intense conversation with a man whose skin was darker than his—a young Saudi meteorologist with a doctorate in cloud physics. The verbal information passed on in a quiet corner had taken his appetite away. The small packet of disks handed under the table continued to torment him—a preliminary analysis of the effects of the Shamal Winds bearing oil-coated particulate throughout the world...detritus made possible by the massive B-52 bombing raids over the sandy reaches of Kuwait and Iraq during Desert Storm. Now, the young man confided, this was merging with an estimated 20,000,000 tons of ash and sulfuric droplets from the eruption of Mount Pinatubo in the Philippines. *The legacy of man*, thought Theodopolus, *and Mother Nature gone mad simultaneously.*

Eyes opening, he reached for the cup. Guessing his contact would approach from the road to his left, the scientist turned in that direction.

4

A "*thump*" against the right side of the aluminum plane took him by surprise. Twisting, pushing back the yoke, he saw a dark shadow fill the starboard window followed by metallic tapping. Bending at the waist, he rubbed a palm-sized patch with his gloved hand and peered out. A slender black object was centered in the clear area. He didn't have time to register the fact that the black Beretta, mounted with a silencer, was held by a soldier in winter regulation U.S. Army dress.

The spot exploded in a searing orange flash. Synaptic junctions exploded in his brain like stringed firecrackers, erratic neuron impulses fired at random into disintegrating gray pulp. The force of the single shot slammed him against the opposite door of the small craft. The small entry hole in his forehead had a counterpart—a jagged, massive exit hole in the rear of his skull now oozing blood, brains and bone chips.

In that micro-moment, the young researcher experienced his last cognitive act; with extreme clarity, his mind's eye saw the proof long sought. Green, blue, yellow, red and black atoms gravitated together, particle-to-particle until a perfect circle of the same color was achieved. Despite the chaos, his understanding was complete. The molecular structure assembling was the embodiment of perfect universal order. Individual color-bound rings moved as if governed by a powerful cosmic force. He saw the impending problem just as clearly. The composite image was also fading away. Theodopolus was privileged, however, before the last of his data-filled cells died, to see them interlock. It was just as he predicted to Walt: resplendent as Joseph's Robe, accelerating with the speed of light, the enjoined rings, truth of the existence of a fifth force–a combination of the four known forces—crowned the Head Frame in a brilliant light and then–disappeared into eternal nothingness.

*   *   *

"IBM management claims there is nothing hush-hush about the work at these labs—other than they don't want their competitors to know how far they've progressed—this according to the guy in the head shed who calls it a, 'Power Visualization System."

"Say again," queried the FBI liaison officer.

"PVS," responded the inter-agency coordinator from Central Intelligence, "is a graphics supercomputer. IBM has think tanks working on this project in San Jose, Yorktown Heights, Zurich, Tokyo and Haifa and . . . ." He let the question fade knowing his counterpart would answer.

"And, the key man—or in the case of Haifa—a woman—have disappeared—all virtually at the same time and without a trace . . . ."

The CIA representative nodded.

Rising, the FBI operative studied the slides of the five missing scientists for a moment then turned to face his peer from across the river and down the road. "There's no apparent tie to National Security so you want us to take over?"

The CIA representative shook his head in the negative.

"My gut feeling is that there is a tie—but I've nothing to go on besides that—and a belly full of pasta that doesn't seem to want to get processed!"

Laughing, the FBI agent from Missing Persons picked up his file. "We'll get started ASAP. I'd invite you to 'take lunch' but sounds like your not a candidate—just yet anyway . . . ."

"Right on Sam. We'll let you take the lead on this but I'll run a slightly slower but parallel track. I'm going to head over to the Defense Analysis Institute. I've got some contacts there that can tell me what's brewing with supercomputers anywhere in the world. If I find anything of interest, I'll call."

Meeting over, transfer complete, the two agents shook hands and left the FBI Conference Room.

*   *   *

Leaning back in the chair customized to his lanky frame, the President rubbed his left leg. Looking forward to an early departure from the White House for a weekend at Camp David with his wife, daughter and one spayed cat, he pressed the buzzer for his appointment secretary. He flipped shut the large folder that contained news clippings, editorials and—cartoons lambasting, mocking and in general—taking shots at his presidency.

A moment later the man was standing before him, folio in hand. Glancing at the leather-bound book, the secretary snapped it shut. "It's juggling time, Sir. Tom Hart is hand-delivering a file he said requires your immediate review. And in one hour there's a meeting in the Situation Room."

Eyes narrowing, leaning slightly forward, the President's brow furrowed, he rubbed his forehead, letting his hand slide over the thick thatch of hair, his hand finally coming to rest on his neck.

"The meeting will consist of yourself, Hart, Yoder, the Joint Chiefs of Staff, Secretaries of State and Defense and, by special invitation, Commandant of the Corps, General Henderson. As soon as the Air Force can deliver Doctor Chen up from a lecture in Dallas, he, too, will be part of the meeting."

Lips pursing, hands palm down on the desk, upper torso now all the way forward, the President asked, "So what in hell is going on?"

The question was interrupted by the ringing of a mud-green phone on the desk. He nodded to the secretary who picked it up and listened in silence, terminating the brief, one-sided conversation with a cryptic, "Send him in." He put the phone back in the cradle. "That's Hart, Mr. President. He's already here. Call when you need me, Sir." The secretary left by the side door as the double doors leading into the Oval Office opened.

Half rising, the President extended his hand to his grim-faced National Security Advisor. Hart, in his mid-forties, balding and developing a slight paunch, shifted the folio from right to left hand.

"I hope you know what you're doing, Tom. Lyman is going shit tacks if I don't attend his lunch meeting." He motioned Hart to a chair.

The NSA shook his head." The Vice President's on his way to that meeting—with your sincere regrets and a weekend invitation to Camp David—at a mutually convenient time. I've got to meet with Yoder, Admiral Montgomery, Generals' Benedetto, Smallzreid and Henderson and get backgrounded for this meeting. Join us as soon as you've gone through this file."

Half-smiling at Hart's adept handling of the crusty head of World Bank, the President accepted the packet across the desk.

"This file, Mr. President, is TOP SECRET UMBRA."

The President rose to a standing position. Mid-November sun filtered through broken scud clouds into the green and gold office. "It's serious?"

Hart nodded. "I've taken the liberty of having your calls held except those coming directly from me. You'll be able to read this file in peace."

Rubbing the skin under his neck, he looked at Hart. "Interesting choice of words." The NSA forced a half-smile as he glanced at the weekly clipping file.

"Whatever's in here must be damned important Tom. Give me a brief overview." The President held the leather packet in both hands as if it contained a deposit from his daughter's cat.

"In the proverbial nutshell, Mr. President—the CIA just learned that security at one of our top secret research installations has been breached. Communications with the facility came to a halt about eight o'clock this morning—Midwestern time."

"The Midwest! Where is this 'top secret facility'?"

"Northern Minnesota, Mr. President—not far from the Canadian border I'm told."

"What's Yoder got squirreled away in the land of ten thousand lakes?" asked the president, remembering his campaign tours into the state.

"A Black Chamber project—sponsored by Chen's Defense Analysis Institute and financed out of discreet long range funds— research on far out 'what if' scenarios. In this case, a contract with University of Minnesota Physics Department. The University had access to an inactive, underground mine on the northern end of the Iron Range. For public consumption," Hart continued, "the work is ostensibly to determine the rate of proton decay—proof which might confirm Einstein's theory of relativity—or so they say . . . ."

"What's the *real* reason for the project, Tom, and *why* have we suddenly lost contact with it?" The President's gray-blue eyes narrowed.

"An investigation is underway to determine what happened Mr. President. As to the program itself, I'd suggest you familiarize yourself with it." Hart turned at the door. "Only twenty pages long, you may find yourself re-reading certain portions . . . ."

Slowly dropping into his chair, the President nodded as his National Security Advisor left his office.

Placing the folder on the desk, the President's fingers drummed the smooth surface. He became conscious again of the familiar pain in his left leg. Looking around the silent office, then towards the rose garden, he saw that the sun had disappeared under thickening clouds. He knew if he rose, went to the window, he could count the protestors milling about the sidewalk, expressing their dissatisfaction with his presidency over a multitude of injustices, real and imagined. Adjusting the reading lamp, he opened the folio. A sheet of textured blue linen notepaper was clipped to the cover page. The short, unsigned message was written in flowing, Spencerian script:

**"If you believe, no explanations are needed.
If you do not believe, no explanations are possible."**

*   *   *

Quinn brought the rod tip up, gauging the distance to the submerged log. A glance at the high-noon sky told him skill and patience wasn't likely to arouse the large mouth bass. Still, he reflected, sports page fishing forecast did say activity would peak at noon . . . .

The recently retired marine officer knew that the recalcitrant bass, dubbed "Brutus," by the locals, was, at best, going to give the hand-tied Leschak lure a half-hearted look.

Weather north of the Laurentian Divide, granite backbone of Minnesota's Iron Range, had been unseasonably warm. Even the natives marveled at how long the trees had retained their colors. The idyllic summer was too quickly coming to an end. Thoughts of wintering in Key West occurred with a frequency in lock step with the falling thermometer. Only the imminent prospect of white tail deer season—one day away—kept him from draining water pipes, covering the woodpile, setting sticky traps in the kitchen and heading south in his vintage Land Rover.

"*Tempus fugit*, Mister Brutus," he intoned, snap rolling the sinking line over still water. The silver streamer plopped into the center of an imaginary circle with authority and the requisite "*glug*." A single twitch brought forth a sudden boil. His spirits soared as the line went tight and rod bent in unison. Lifting the rod overhead in a quick, smooth motion, he set the barbless hook, anticipating the large mouth's instinctive lunge toward the sunken log—wrap-around and break off— Brutus's oft-used survival tactic.

"Gottcha!" muttered Quinn to himself.

9

Denied safety by entanglement, Brutus, with one powerful beat of his tail, reversed course and instinctively headed for deep water.

Quinn swung around to keep fish, rod and body aligned. He was alone on the lake, summer residents having returned to Hibbing or points further south. Only a bald eagle soaring above shared the mid-day hour along with an occasional loon, wings beating rapidly, its thoughts on Florida migration as well.

Cupping the reel, the line slipped away until yellow backing appeared. Lifting the rod tip higher, he applied maximum pressure. The bass exacted a measure of revenge as cold water poured in over the tops of the waders. Sucking in his breath as the water cascaded past his private parts, Quinn backed toward shore slowly, repurchasing lost line as he shuffled over the sandy bottom slippery with a layer of red and gold leaves. He was already anticipating the sight of legendary bass as he brought it to submission and then its release.

"Nice going Colonel."

Startled at the soft voice, his head snapped around and the tip dropped slightly. Instantly the line went limp. Tension gone, the battle was over. Turning away, brow furrowed, Quinn barrel-rolled the flyline forward in mock frustration. When he turned again, he pursed his lips to keep the smile from spreading, but his eyes, crinkling in the corners, gave him away.

Karin Maki, slim, dimpled and grinning, in jeans and sneakers, stood on the end of the dock, hands tucked in hip pockets, pale blue eyes, a clue to her Finnish heritage. The eyes matched the powder blue down vest worn over a white-hooded sweatshirt proclaiming the virtues of the Riverside Inn and it's succulent burgers. She stood framed against birches, red maples and a melange of white and Norway pines. Quinn appreciated the natural beauty of both woman and woods.

"What brings you out today?" asked Quinn slogging through the water, reeling in line.

Karin rolled her eyes theatrically, "Thursdays Colonel Quinn—I come every Thursday and clean your wretched mess hovel."

Laughing, he tapped the side of his head. "Memory Ms. Maki, first function to fade." He extended a hand to his comely housekeeper. "Give an old grunt a hand—it's colder'n a witch's wigwam inside these waders."

Karin's hand came out and she braced herself. "Grandma sent soup and sandwiches. She didn't think you were eating regular meals. Come on!" Quinn came out of the winter chilling lake onto the dock.

He watched as the young woman, recently separated and the mother of a twelve-year old daughter, headed back up the path to his cabin.

Quinn's near-monastic retreat to the northern reaches of Minnesota, was a home-brewed prescription for recovery from the trauma of a twenty-two year marriage gone sour and an equally long career in the Marine Corps brought to an end through request for early retirement. He found himself at ease in the company of the young woman but the memory of his ex-wife, faced covered with a black mourner's veil was never far removed.

Sitting on the overturned boat, now beached for the winter, he eased off the waders. A sudden breeze rippled the water. Looking west, he saw the gray clouds forming beyond the tree line. Waders over one arm, rod in the other hand, he turned toward the cabin. Half way up the slope, he saw Karin come out on the cantilevered deck, remote phone in hand. "You've got a call!" Her voice echoed over the small bay behind him.

Quinn raised the rod in mock salute. "I'll call 'em back," he shouted. A moment passed before Karin returned, cradling the phone between her breasts.

"It's a General—in the marines. He want's to talk now! He says he won't take a call back."

Quinn snorted, laying the fly rod on a picnic table, "Toss me the phone. We'll see who's giving out marching orders!" He grinned up at Karin as she lowered the phone by its antenna tip.

*   *   *

Quinn spooned down Grandma Maki's split pea soup, staring out the pine framed bay window at a muskrat forging a "V" in the leaf-flecked waters of the kettle lake.

"You haven't said a word since coming out of the shower, Caleb. Did I handle that call wrong? Should I have taken a message instead." Karin's spoon hung in the air.

Quinn broke off the thousand-yard stare and turned toward his young housekeeper, a soft smile on his face. "Sorry Karin. No, no, you handled it just fine. That was General Jack Henderson, Commandant of the Corps." Quinn chuckled. "We owe each other beau coups favors and '*Jocko*' just called in "*a wee small marker.*"

*   *   *

Karin watched Quinn standing on the balcony, his hand outstretched, feeding a small flock of redpolls that hopped from pine-bough to feeding station and back. Seeing the relaxed smile on his face, she resolved not to scold for letting birdseed spill inside the cabin. The small black objects, she'd said, made it impossible to tell if mice had taken over the cabin.

She found herself wondering how the one-day-a-week job could be made to last beyond the summer. Knowing he planned to close the cabin in two weeks and head south with the close of the white-tail deer season left her feeling melancholy. Separated, she was torn between reconciliation or removing herself from the marriage. Thoughts about the retired marine who took her daughter fishing and taught them both the art and science of making "*GitMo*" burgers occupied a growing amount of her time. Her daughter had asked more than once if she was 'interested" in the "*good looking guy*" but Karin had put off the query with a blush and laughter. She did think about the "*guy*" and wasn't so concerned about any differences in age as the fact that his cabin was the equivalent of a small town library–books of all sorts of subjects could be found everywhere. When he wasn't hunting or fishing, Karin had noted, he was reading. Intellectual disparity–real or imagined–was her greatest problem. If she could come to grips with that, she thought, *perhaps there was a chance* . . . .

Quinn turned and smiled, pointing to the array of shotguns stacked against the daybed couch. "Having a job to do doesn't mean we can't enjoy. Karin! To hell with housework! Let's do some road hunting on the back roads going up to Tower-Soudan. Maybe we'll even have time to hit the gravel pits behind Ely. They tell me the grouse are thick as ticks up there."

"Tower-Soudan? You've got a job there—for a general in Washington?" Karin cocked her head.

"I've got to get a first-hand report on a project and relay the information back to Washington. That's it! Jocko assured me it'd take a couple of hours at most. I get on the phone and *finito no mas*! We get

12

some hunting in and—I'll take you to dinner at the Silver Rapids Inn." He held up her jacket in invitation.

Karin hesitated but her gaze went to the guns.

"Take your pick," he said, "and we're gone!"

"You're serious?"

"Very," Quinn responded with mock solemnity. "Time to take a break. As I recall, Tammy's with her dad for the weekend and the dust balls can clone unobserved while we're gone. Besides, I'm anxious to see this Finnish Annie Oakley they talk about at the Riverside in action! Loosen up Karin, smell the gunpowder along the way!"

Karin looked again at the guns and back to Quinn. "I'll have to cut class."

"Class? What class?"

"Med Tech. Always wanted to be a nurse. Townie's got to be Candy Stripers—I'd be swilling pigs—they'd be cleaning bed pans—I envied them."

Grinning, Quinn shook his head.

Karin, frowned, then laughed. "You're right. It's time to smell the gunpowder."

Choosing a 20-gauge pump, he thrust it in her arms. "Try this on for size."

Karin cleared the chamber and drew a bead on imaginary grouse winging beyond the balcony. "Bang! Bang!" she shouted. "Two shots—two birds. How's that?"

"Second rule of the old Camp Matthews rifle range: You never miss when you're dry firing," deadpanned Quinn with a twinkle in his eyes.

*   *   *

"Impressive! Two shots—three birds." Quinn gave Karin a thumbs-up sign as the Rover moved eastward. Ravens worked the highway for road-kill remnants. Blue Jays and gray-black Loggerhead Shrikes darted into view, then disappeared in the pines. Driver and passenger shared black coffee from a heavy porcelain cup.

"One-ten, Caleb. If you spend any time at all with your contact, we're out of hunting time."

"You're right. But we've got four birds. Presented cleaned, plucked and with some Irish blarney at the Rapids, we can count on partridge pie with wild rice and hazelnut stuffing—a Caleb's special and voila—dinner *tres magnifique!*"

"Partridge pie sounds wonderful but a Caleb Special?" Karin took the mug back, reaching down for the thermos.

"Ah," mused Quinn, "the Special. It takes some describing but what the . . . ."

Startled by the sudden change in his tone, Karin looked up.

Releasing his foot from the gas pedal, he let the British army-green four-wheeler come to a slow stop at the "T" junction. Northbound access was blocked by an army jeep and a white armored personnel carrier, its top cluttered with antennas. All units, stopped and moving, were in winter camouflage—white over summer jungle patterns. A young soldier in winter dress, metal helmet and armed with an M16 rifle stood between them and the road leading to Tower-Soudan. Braking to a complete stop, Quinn saw three uniformed soldiers in the jeep.

Beyond them—rolling north—a convoy was moving at a slow but steady pace. The noise enveloped the metal cab of the Land Rover. A line of Army trucks carrying soldiers were followed by tanks and fuel trucks. He recognized the tanks as vintage M48's, tracks rubber-cleated for over-the-road transport. The drivers wore Arctic masks and goggles. Armored Personnel Carriers—APC's—were interspersed and carried .50 caliber machine-guns. Instinctively, he counted the units—twelve tanks, seven troop carriers—wondering if he'd seen any of the vehicles in their Vietnam prime. The sight brought back twin memories—elation and fear—impending action supplanting grinding boredom.

"What's happening?" asked Karin.

"Must be weekend maneuvers," he said. Something about the convoy struck him as unusual but he couldn't put his finger on it.

Karin leaned forward, looking at the sullen-faced soldier staring at them.

"It must be the National Guard. The guy with the gun—he's a friend of Matt's. He belongs to a unit stationed at Virginia." She turned. "Hah! And I know why he's not smiling."

Quinn glanced in her direction.

"Deer hunting season starts—Saturday!"

14

Quinn nodded. "I'll tell you what," he said, "you talk with Matt's friend and I'll nail the officers sitting in the jeep. We'll compare notes and see what we've learned." He didn't disclose his nagging thoughts of a connection between General Henderson's call and the white convoy headed in the same direction he'd been instructed to go.

Quinn was halfway to the jeep when the rider in the front spotted his approach and swung out of the enclosed vehicle. He stared at the young man in battle gear, white vapor barrier boots and a regulation 9.mm pistol strapped to his webbed belt.

"Captain Quinn, it's me, Sergeant . . . make that Captain Holmes."

Quinn took the redheaded, freckle-faced man's hand, a look of surprise on his face. "Sergeant Holmes, Ernie Holmes, First Recon— Khe Sahn. I don't believe it!"

The Guard captain flashed the grin Quinn remembered well. "It's me all right. Still at it, but in the Guards—putting the checks into my daughter's college fund. What are you doing up here? I thought you'd be running the Corps by this time."

Quinn smiled, pulling the collar of his thermal-hunting parka up against the dry cold. "Not in the cards, Ernie. I cashed out with twenty-two and eagles."

Holmes smiled. "Congratulations on the big bird."

"Thanks, Ernie but tell me, war games on Friday? Opening of rifle season Saturday? That's damn near sacrilegious in this neck of the woods. What gives?"

The Guard captain jerked his head in the direction of the jeep, voice dropping. "You nailed it Colonel! At nine this morning I was giving a math quiz to my ninth graders and . . . well, look at me now!"

Watching a tank churn by, Quinn turned to his former squad leader. "What's your assignment on this maneuver?"

"Official bag man," responded Holmes, green eyes laughing.

"Bag man? Like the one with money?" asked Quinn with a smile.

Holmes nodded. "As good as money. When our rolling stock is on the road, you can count on a lot of minor damage claims—broken mailboxes, fields torn up—you name it! I'm Community Liaison Officer. If a problem comes up and money can solve it, I'm empowered to issue a voucher good for up to two hundred dollars—on the spot—

15

sign the waiver—you gets the money! In addition, I'll take my turn as perimeter defense officer once we reach our objective."

Quinn grinned. "Maybe I can get you to scrape an APC-against the Land Rover back there. It needs a new paint job." He was glad his former re-con man mentioned the word "*objective*."

The two men laughed.

He pointed toward the armored personnel carrier idling behind the jeep. Four soldiers stood around the APC, M16's in hand. "What's with the big antenna—Guard got something the marine's haven't seen yet?"

Looking over his shoulder, first at the APC, then to the jeep, Holmes faced Quinn, smile gone, voice subdued. "I don't have the foggiest notion, Colonel. They don't want me to talk to the crew. Those troopers around the rig are tough hombres to boot!"

"*They* are?"

"Regular Army. The guy in the back of the jeep—He's a full bird colonel but not real communicative. The carrier is his private command track. The RA's kicked our crew out, then brought in all sorts of black box gear and mounted the antenna themselves—they've got another crated inside the rig. Between that and the fact we're carrying a full complement of live ammo tells me this is one serious maneuver. One of the first guards to report said he saw them loading small crates as well. Didn't know what they were."

Quinn whistled quietly, pointing at a tank rolling by.

"That M48 is loaded with live rounds?"

"Sixty-three rounds per tank," responded Holmes. "Full ammo boxes for the M60D' machine guns and six, twenty-round clips for each M3 grease gun. We've got armor piercing and HEAT rounds on board. The whole megillah Colonel! Kinda'of a surprise about how that came about, too."

Quinn cocked his head.

"Our units came out of Hibbing, Chisholm, Eveleth and Duluth in summer colors. Normally, we'd have a marry-up at the armory in Virginia if we were headed north. This time, the marry-up occurred on top of the Laurentian Divide—five miles north of Virginia at the tourist lookout. Ammo trucks from Camp Ripley were there with the ordnance and 50-pound sacks of lime. We had to swab down the tanks and carriers—German panzer-style—as the ammunition was loaded. It was damned unusual!"

"Ernie—you remember convoys in the Corps?" Quinn finally focused on what was bothering him about the convoy.

Holmes nodded, the familiar grin returning. "Saw a lot of 'Nam from the top of carriers!"

Quinn pointed at a spectral tank clanking by, "Unless we expected a firefight around the corner, TC's were ordered to swing the turrets to the rear and lock the barrel in the road-dolly for safety purposes. Does the guard have a different SOP?"

Holmes made a short jerking motion with his hand. "Orders from Snowtop in the back seat, Colonel. Everything is to be treated as if we were under combat conditions. Oh, there's something else. Each tank carries a full complement of APER's."

Suppressing a shudder, Quinn remembered all too well anti-personnel shells—each round packed with 5000 razor-sharp flechettes. He'd seen Viet Cong and NVA soldiers literally nailed to trees by the deadly weapon—his own men cut to shreds when a round became 'friendly fire'. He stared at Holmes. "You're going into deep bush, I hope—away from cows, pigs and people?"

Holmes shook his head from side-to-side. "One more thing, there may be more people going on this weekender—one of the APC's seems to have personnel aboard who have been heard but not seen. Its the same one that's got a half dozen crates about five foot long, maybe two feet high. Nobody recognized them—painted black—no markings . . . ."

"Black boxes, six you . . . ."

"Captain Holmes!" the order barked out from the jeep. The Captain spun around. He turned back briefly. "I'll be back in a minute. Snowtop's side-kick beckons." Holmes retreated toward the enclosed jeep. Glancing over his shoulder, Quinn saw Karin still engaged in conversation with the road guard. More vehicles had piled up behind the Rover. One was a heavily loaded logging truck, sixty-foot aspen tips bending to the surface. The driver had dismounted and was heading toward the road guard.

"Colonel Quinn."

He spun around at the sound of the unfamiliar voice. A tall, square-face Colonel with piercing black eyes and close-cut white hair showing under his field cap stood before him. Glancing over his shoulder at the jeep, he saw Holmes sitting stiffly in the right-hand seat, looking straight ahead. Quinn suspected that his former squad leader had been dressed down for talking to a civilian. "It's Caleb Quinn, Colonel, retired." He studied the regular army officer. *Squared away . . . looks like a Drill Instructor . . . eats nails for breakfast . . . shits'em for supper.*

"Sorry for the inconvenience, Colonel Quinn. The convoys' just about past. Captain Holmes says he served under you in Vietnam. I'm curious, what brings a retired marine officer up to this God-forsaken country?"

Flat voice. Doesn't sound like it comes from anywhere in particular. No twang, nothing nasal, like a mechanical toy. "Hunting and fishing," Quinn replied, "and you're just a few miles south of the Boundary Water Canoe Area. You've heard of it?"

The man was silent, as if formulating an answer.

"Where are you from?" asked Quinn casually, hoping to get back to the issue of the Guard's objective. Shrugging his shoulders, the army officer replied; "Fort Hood, assigned to armored command training school."

"No, I meant originally—where's home?"

"Ah!" responded the tall man. "I misunderstood. I'm from Pennsylvania—Pittsburgh to be exact."

Like hell you're from Pittsburgh was Quinn's instinctive reaction. "Your name? I've introduced myself," pointing toward the blank left-hand chest pocket of the officer's battle jacket.

"My apologies, Colonel. We flew up here on commercial airlines. I'm using gear taken from the armory in Virginia. My name's Hartlett—Colonel Charles Hartlett, Regular Army." He extended his gloved hand.

Surprised that he didn't remove the glove, Quinn accepted it and felt the vise-like grip.

A deep voice boomed behind his back.

"Colonel Hartlett! Citizen Lubokowski here! That's my lumber rig you and your toy soldiers have got stalled back there. I'm asking you to

18

haul ass! Fuel costs money and deer season opens this weekend! I wanna move!"

Moving past Quinn, the wide-shouldered driver stood nose-to-nose with the regular army colonel.

Quinn could see the officer's eyes narrow.

Hartlett's voice was steely. "As I was telling your fellow *citizen*, the convoy's rear guard is about to pass. You can get your . . . your lumber wagon on the road when it clears!" He spat out the words, turned on his heel and headed back to his jeep.

The logger smiled at Quinn. "You hear that crap . . . my lumber wagon . . ." The burly man slapped Quinn on the back and headed toward his rig. "Lumber wagon . . . I'll be dammed . . ." he mumbled, shuffling toward his idling rig.

Quinn saw the command jeep pulling in behind the duce-and-a-half carrying the black-on-white sign reading, "END OF CONVOY." Holmes did not look back.

Glancing at Karin, already in the Land Rover. She responded with a smile. Snow flakes peppered his face as he made his way back, wondering where Hartlett was really from—and why the need for deception. Experience with men from every part of the country gave him a knack for picking out their home state before they volunteered it. Sliding onto the bench seat, he glanced out the window adding the people and black boxes to the jumbled scenario. He looked to Karin—"So much for precipitation beginning late tonight."

Karin laughed and poured coffee into the communal cup.

Quinn started the Land Rover, then pulled in behind the last vehicle in the convoy. They drove in silence. He took the coffee mug and glanced at the speedometer—thirty miles an hour in the convoy's trail. He was thinking about the regular army colonel's appearance, demeanor, voice and the word, "lumber wagon." He turned toward Karin, "What did you learn back there?"

"Total surprise to the Guards. Ollie got his call at Hibbing Taconite at ten thirty-five. They gave him three hours to get to the Virginia Armory with his gear. He was told that the operation was scheduled to last until midnight Sunday night."

19

Quinn rubbed his chin. "Did this Ollie say anything about the maneuver itself? What's the plan? Who's involved besides tank units and regular army observers?"

Karin had slipped off her Nikes and was busy lacing up the fur-topped Sorrels. She had already slipped into the snowmobile suit stuffed in the back of the Land Rover. She half-turned. "Whatever the plan is, it's in the Tower-Soudan area. Ollie said the rumor mill had them securing a specific site and defending it against aggressors. All units were called out. That's all he knew."

Objective is in Tower-Soudan. That makes two up there. He shook his head. "Nothing more uh?" He looked quizzically at Karin.

She rubbed her forehead for a minute. "Yeah! Ollie said this is the first maneuver he's been on with so many regular army involved."

"How many is many?"

"Twenty that he could count. Each tank was to have at least one member of the regular army on board. Communications vehicle was all regular army."

"Sure there's nothing more?" Quinn pressed Karin. She stared at the thickening snowflakes plastering on the windscreen, melting and being shoved aside with the wiper blades.

Suddenly, she reached for his arm resting on the gearshift lever. "Ollie said in the past, regular army guys would talk up the maneuver—they always knew what was coming. These guys were like clams. No conversation, nothing. And, what ever is in that one big vehicle with the antenna must be something special."

Quinn glanced at Karin. "How so?"

"Ollie thinks there are some people in there. He heard one loud voice and started for the unit but was headed off by a regular army type."

Head shaking, Quinn stared at the thickening snow.

The word "lumber wagon" came back into his mind—as did the regular Army colonel, wide-set eyes and flat speech, General Henderson's call and the man he was to meet—everything coming together in the Tower-Soudan area! *How does it all tie together?*

Braking to avoid hitting the last truck in the convoy, Quinn turned his left signal at the arrow pointing to Hoodoo Point Park on the western edge of Tower. As they left the main road, heading for the

20

grass airstrip on the peninsula jutting into Lake Vermillion, the white convoy disappeared into the blowing snow.

*   *   *

Quinn glanced at his watch. "Not even four o'clock and it's going to be dark in half an hour."

"You're meeting someone out here?" asked Karin, staring over the hood of the Rover at the row of small planes tied down on the grass strip.

"That's what Jocko said. My contact is going to be working on a Luscombe at the far end of the line. I'm to walk up, identify myself, get the man's information, then find a phone and relay it back to Washington." Quinn slipped on his thermal gloves and reached up past his hunting bow suspended beneath the roof of the Land Rover, pulling out a black woolen watch cap. He adjusted it around his ears, took a flashlight from the open glove compartment and turned to his hunting partner of one day. "It's what well-dressed messengers are wearing this season."

She laughed. "It's getting cold out there. Do you think you'll be long?"

"Minutes, Karin—just minutes."

Falling snow had increased from small, dry beads to larger flakes that melted on the engine's hood but started to mound on the fenders.

Quinn trudged on, making mental notes of the aircraft he passed— vintage Cubs, an ancient Aeronca and an occasional Cessna. Wind pushing at his back, darkness seemed to leach out of the cold ground.

The designated plane was tied down sixty feet away. Wind made the small metal plane quiver. He guessed over an_inch of snow had fallen since the enforced stop at the junction. Scanning the ground for tracks—he saw nothing to indicate the presence of an earlier arrival.

Jocko, I hope to hell you didn't send me on a wild goose chase. Where is this son-of-a-bitch with something to say? Flicking on the flashlight, Quinn let it play over the two-seater. *N4781G—Golf. Right plane, right place. Damn!"* He glanced in the direction of the Land Rover. It was nearly lost in the falling snow and darkness. Turning back, he let the light play over the small aircraft. The left-hand door was slightly ajar.

Somebody's been here. But why leave the door open?

21

Keeping the beam of light on the door, Quinn approached the craft, ducking under the wing. Then he saw the thin, dark strain of blood running from the bottom of the door, down the outer shell into the snow. He could see it had already congealed. He touched the door with his gloved hand and then had to quickly step back as the door pushed open from the inside.

The body half fell from the aircraft. The bare index finger of the dead man grazed the puddle of wine-dark fluid. The pointing finger sent a chill through Quinn's body. The back of the man's skull was gone.

Sweet Mother of Jesus! He stared at the young man's face from the side then knelt to examine the wound in the forehead. *Small caliber, very close range. Who ever got the poor guy must have stood right next to the plane, held the pistol to the window and fired once.*

Quinn glanced around. Darkness enveloped the field. The flakes seemed to have gotten bigger. He stood up, letting the flashlight's beam play over the inside of the aircraft. Nothing looked like it had been touched.

He lifted the man to a sitting position and removed his gloves.

So young. So many things left undone. The familiar litany from post-battle letters raced through his mind. Kyle's face appeared momentarily then faded from view. Mumbling the prayers he'd learned as an altar boy, Quinn traced the sigh of the cross on the man's forehead—a ritual he'd repeated many times before.

He quickly felt inside the pockets of the man's parka but found nothing. The dead pilot stared with unseeing eyes. On a battlefield impulse, he opened the man's shirt, exposing a gold chain. He pulled it up bringing an intricately designed cross into view. He unhooked the chain and pulled it from around the man's neck. *A Dog Tag of sorts*, thought Quinn . . . .

Zero, zip! Not a clue! He must have thought it was me coming to say the magic words so he just sat there.

Quinn backed away from the door, pocketed the cross and chain and squatted on his haunches. *What now, General? Whatever the message is, it's right there.* Quinn ran out of ideas as the light glistened on the dark glob.

22

He pushed the door shut and backed away, switched off the flashlight and headed toward Karin and the Rover. *I'll call Jocko, but I don't know what to tell you.*

*   *   *

Her hands wrapped tightly around the porcelain cup, Karin shuddered as he described the scene at the end of the runway. Vapors from the open thermos misted the windows.

"What'll you do, Caleb?"

Shifting slightly, he continued to stare out into the night. "I'll call the number, Karin. I'll tell Henderson what happened. If I know him at all he'll pick my brains and then . . . ." The words faded away.

"The body Caleb. What about it?"

Quinn's eyes narrowed for a moment and he faced Karin. "The man was murdered. If I turn the word in, there are more questions to be asked than you and I have answers for. My footprints are back there. God knows what I might have left in the way of finger prints."

Karin tapped his thermal gloves with her fingers. "You were wearing these all the time?"

"Nice try, Karin, no. When I searched his pockets I took them off."

"Find anything?"

Fumbling in his pockets, he handed her the cross and chain. "What do you make of this?"

She let the cross dangle in the dim light. It twisted slowly for a second, then she saw the initials. "He must have been Greek, Caleb. This is a Coptic cross . . . ."

Suddenly, the night sky lit up in an intense white glow. A second later the sound and shock wave hit the Land Rover.

"Jesus!" Instinctively Quinn grabbed for Karin, pushing her onto the bench seat, his body over hers. He waited until he was sure there wasn't going to be a secondary explosion, then released his protective hold. At the end of the line, what had once been an aluminum, two-seat aircraft, burned furiously in an intense yellow blaze. It created a hallo effect in the falling snow and advancing darkness.

Quinn glanced at his watch. It was 4:31 P.M.

Swearing under his breath, he put the Rover into reverse, backing into a tight turn then slammed into first and headed down the road

23

leading back to the Village of Tower, lights off. "Damn Karin! We're gonna get outta here before the fire crew and police see us!"

Wind driven snow swirled around them. Quinn tried to guess at the reaction time of a small-town volunteer fire department. He turned to Karin. "Keep your eyes open for red lights ahead. As soon as you . . ."

"Now Caleb! I seen lights now!" Karin pointed down the narrow road.

He quickly swung the Rover off the tarred road to the left. Climbing a slight grade, he snapped the headlights on—then off. They reflected off polished marble and granite tombstones. Circling to face the way they'd come, Quinn pulled under the eaves of a casket unloading overhang and slid his window open. Seconds later, the first of the volunteers roared past. Private cars, pick-up trucks with roll bars and a fire truck that looked as if it had seen military duty, headed toward the small sod airstrip.

Quinn was silent, thinking about his next move.

"So much for fingerprints, Caleb," Karin said quietly.

Nodding his head, he shot a sidelong glance at the young woman who seemed to be in step with his own private thoughts.

"That's good news—for a while. That fire wasn't accidental, Karin. Aviation fuel doesn't burn like that. Whoever killed our Greek friend left a timed explosion—I'd guess set for four-thirty."

Karin quickly glanced at her watch and then reached for his hand. "You were standing next to the plane at 4:25."

He squeezed her hand. "What concerns me more is getting the hell out of here."

Quinn waited until they'd seen the last of the fire fighters. Lights still off, he pulled onto the main road. "Until we can find a phone let's hope and pray nobody pays any attention to us and that they trample the Land Rover tracks and my foot prints."

"I'll pray in Finnish Caleb. Great grandmother Palo told me it was twice as effective." Quinn laughed softly.

"By the way Colonel Quinn—does this mean dinner *tres magnifique* just went out the window?"

He gripped her forearm "out the window but not forgotten. When we get out this bottomless septic tank, I'll make it up in spades."

24

Karin kept her eyes on the main street of Tower ahead and a tight hold on his hand.

*   *   *

He pulled to a stop directly across from the small cafe with curtained windows. The main street of Tower was deserted, one snow-covered car parked in front of the only open restaurant.

"The plan Karin . . . you ready for this?"

She nodded.

"You order supper—I'll get on the phone. While I try to reach Washington, ask about the maneuvers. Pick up whatever you can. Maybe I'm still too close to the military but there's got to be a link between Henderson's call, the dead man and the tank operation."

Karin nodded.

Quinn made a U-turn in the wide street.

He parked and locked the Land Rover.

"Afraid somebody's going to rip you off?"

Turning the key on the right-hand door, he laughed, "Karin, half my net worth is in this rig. Between deer rifles, ammo, bow hunting equipment, some fishing and cold weather gear, I'm a sports shop on four wheels."

The wind whistled down the wide streets of the former mining community as Karin and Quinn entered the cafe. One elderly couple occupied the table closest to the counter. They looked up, gave the outsiders a casual glance and resumed eating. A woman, in her early fifties, came out from the back room, menus in hand. The phone was in the small passageway between the outer door and the one leading into the restaurant. He motioned to the table closest to the window and headed back for the phone. "Order me anything hot," he said on his way out.

Fumbling in his pockets, he found the eleven-digit sequence scribbled on the back of the bank deposit slip. Punching numbers, he shifted time zones one hour, wondering where the number he was ringing would end up. A woman's curt voice answered on the second ring. "You've reached 7750. May I have the number you're calling from?"

25

Quinn glanced at the number, barely legible in the dark and gave it to the operator. The voice at the other end responded: "Please hang up and stand by. Your call will be returned momentarily."

He quickly interjected. "Miss! I'm calling from a pay phone!"

The response was pleasant but officious; "You're call *will* be returned, Sir!" The phone went dead.

A cold blast greeted Quinn. He turned to see the door to the outside open and a swirl of snow filter into the small hallway.

A thick-necked, stocky man, a half-head shorter than himself, filled the open door with his bulky frame. The star on the flap of the beaver-trimmed Russian-style hat read, "Deputy Sheriff." His voice was brusque. "You own that four-wheeler out there?"

"That's right. Any problems?" responded Quinn.

The squat man's right hand dropped onto the grip of his holstered pistol. Quinn's gaze followed and recognized the non-standard .357 magnum. Experience told him which men with guns knew what to do with them—and which ones didn't. He put the stocky law enforcement officer in the former category.

"Been in town long?" queried the police officer.

Quinn turned, shoulder resting against the phone he hoped wouldn't ring. "Just got here. Been road hunting but got blown out by the weather."

The man rubbed at his runny nose with the back of his gloved hand.

"Where might home be?" he asked, right hand still caressing the grip of his holstered gun.

Quinn held his response in check, glad the conversation was taking place in a shadowed area. He knew his face was growing darker. "Home, Sheriff, is on Beatrice Lake."

The young man's eyes flickered momentarily. Quinn knew living in the area was worth something.

"Your driver's license—please." The man's hand went out, palm up.

Quinn reached in the pocket of his down hunting jacket. As he pulled the wallet out, the gold cross and chain fell to the wet floor. Eyes on Quinn, the police officer stooped down and retrieved it.

26

"Must be worth a fair piece of change Mister. Shouldn't it be around your neck?"

"Belongs to a friend." He regretted that he hadn't left it in the Rover.

Searching for his license, Quinn was aware of another impending problem—his change of address from Florida to Minnesota. Handing the plastic card over, he accepted the cross and chain back but noticed the deputy had turned the cross around. The deputy was pulling his flashlight from his black service belt when the phone shattered the silence in the small, cold hallway.

"Shit!" The words came from Quinn's lips involuntarily.

The lawman looked up. "You expecting a call? Go ahead. I can wait."

He lifted the receiver from the hook. "Quinn here."

The raspy voice of the Commandant rang in his ears. "Colonel Caleb—no middle initial—Quinn, USMC, Retired?"

"Correct," he responded.

"Military service number and date of retirement?"

Hands sweating, he knew Henderson was doing what he had to do. He also knew the sheriff was watching him intently.

"Sorry, this connection isn't working. I'll replace the call. Hang on—I'll get right back." He put the phone down before Henderson could respond and turned to face the young lawman.

The eyes of the deputy narrowed. "Nobody here but you and me. And I don't need to use the phone unless to check if this is really yours." He held up the card, letting the light play on the picture. "You know something, Mister? I've got you pegged for at least two violations of Minnesota vehicular law and I'll bet if I look a little further, I could nail you on some more. Open gun cases, loaded weapons and all those other little goodies we get to handle this time of the year." A thin smile flickered across the young man's face.

"Two violations sheriff? I may be overdue to get a Minnesota driver's license but I'm still registered in the State of Florida. What's number two?" Quinn's voice took on a steely edge.

The deputy sheriff kept his eyes on Quinn and jerked his head over his shoulder. "You screwed up there. Making an illegal U-turn on the main street of Tower. Gottcha' dead to rights and didn't even have to see you make the turn. Judge Palo'll buy into the charge."

Quinn threw up his hands in exaggerated frustration. "Guilty as charged. Give me a ticket."

The sheriff handed Quinn the license. The tone suddenly changed.

"No Mister Quinn, I don't want you coming up here, leaving your hard-earned dollars then going to Florida with a bad taste in your mouth. I just want to ask a couple of questions about your vehicle— where it's been the last hour." The young deputy paused, "see, we gotta problem. A plane just burned on HooDoo Point. And maybe there was somebody in it. Volunteers recall seeing a single set of tracks coming out but drove all over'em getting to the plane itself."

Quinn held his breath. *Jocko Henderson, damn your ass. A wee small favor and a couple of hours because I'm the closest warm body to your problem—whatever in hell it is.*

The door from the restaurant side opened throwing a field of light on the two men. Karin froze in the doorway. She looked at Quinn, the sheriff and back to Quinn.

"She with you?"

He nodded. "My hunting partner."

"She with you down on Hoodoo Point?"

Quinn interjected quickly. "We weren't on HooDoo Point Sheriff. Never heard of the place and a damned strange name for an airport."

The deputy looked at Karin. "You must be from around here. His gaze dropped to her ring finger.

"I'm Karin Maki from Side Lake. My dad's originally from around here."

"And who might your daddy be Miss—or is it Mrs. Maki?"

Quinn could hear the steely bristle in Karin's voice. "It's *Ms.* Maki and my *father's* name is Frank Maturi!"

The deputy's eyes opened wide.

"The head of HibTac Union—Big Frank Maturi?"

"One and the same."

The heavy air in the small enclosure lightened. Quinn made a mental note to ask about Karin's father.

28

"No tickets, Mr. Quinn. And if *MS*. Maki says you weren't at HooDoo Point, why, that's good enough for me. Enjoy your supper here at the Tower Cafe. It's the finest Finnish fare money can buy."

The sheriff back-pedaled out of the doorway into the wind driven snow.

Quinn grabbed Karin's arm. He spoke under his breath after the outer door closed. "This is starting to get pretty hairy. I want you to get the hell out of here! This is like being in a room full of flypaper—everything I land on turns to sticky shit!" He stopped just inside the restaurant and lowered his voice. "Do you have relatives, friends—anybody you can stay with here in Tower—better yet, anybody who can drive you back to Side Lake tonight?"

Karin shook her head grinning. "You're stuck with the Finnish Annie Oakley, Colonel. Besides, word on the cafe radio is we're gonna' get dumped on. The *'light precipitation'* is turning into a snowstorm, possibly a blizzard. I wouldn't ask anybody to take me home."

He stared at her in the dark hallway. "I'll say it again, Karin. This situation is sliding out of control. I won't BS you, I'm getting nervous." Quinn looked through the small square window on the outer door. He reached for her arm, pointing through the frosty glass.

Flashlight in hand, the deputy was squatting in the swirling snow behind the Rover—examining the track pattern. Quinn shook his head in disgust.

"I should've taken those field tires off. If they find so much as a six inches imprint they'll know we were within a block of the damned plane."

Karin loosened Quinn's grip on her arm, a smile flashed across her face. "Make your call and come back in. I'm going to have a waitress serve tonight's special."

Quinn looked back to see the red taillights of the deputy's car fade into the blowing snow.

"And one more bit of military intelligence, Colonel Caleb Quinn. About the Guard maneuvers—the lady who runs this place said one of the jeeps stopped here to make a call and she overheard the men talking about where they were headed. From what she could make out, it's the Tower-Soudan mine."

*   *   *

Quinn pushed the buttons on the wall phone. It was answered on the first ring. A man's voice completed the drill but he was not asked to hang up. A moment passed and Jocko's familiar, gravelly voice was back on the line. "Caleb, is that you?"

"Yes sir! Colonel Caleb-no middle-initial-Quinn, USMC retired. Service number 01620409. Sorry about the abrupt . . . ."

Henderson cut him off. "Listen, I'm on a secure line but that doesn't do either of us a hellavu'lot of good because you're not. I'm going to start a scramble. Hang ten, button's being pressed now!"

Quinn counted ten seconds.

"Okay, let's talk," said the General. "Did you make contact? Let's have the message."

Looking down at his hunting boots and the puddle of melted snow at his feet, the phone was cold in his grip. "No message, Jocko. My contact was dead on arrival. He apparently took a small bore shot to the forehead at about three-thirty—my time—in the plane designated. No sign of his assailant. We're in the leading edge of heavy-duty snowstorm! Minutes after leaving the scene, the Luscombe blew up. Best guess is that a timed thermite device was dropped in the fuel tank. Little or nothing remains except . . . ."

"Except what?" Henderson's voice was subdued.

"No papers, wallet rings. But I did take a gold chain and cross from his neck. Force of habit I guess—collecting Dog Tags. A friend with me identified the cross as being Coptic—Greek Orthodox—it has the initials "TT" on the back. My hunting partner thinks the man was Greek because of the cross."

There was silence at the other end of the line.

"Is this friend of yours with you now?"

"She's in the restaurant."

"She? Fry pan to fire and back again?"

Quinn laughed. "Not that! She's my housekeeper and right now, one very nervous hunting partner."

"For a housekeeper, Caleb, she's pretty astute. Your contact's name was Theodore Michael Theodopolus, a scientist who worked for the Company. What he had to tell you was vital. With him out of the picture, we've got a major problem."

30

Quinn coughed. *"We?* What's with the Royal *'We'* Jocko? I retired from the Corps—not the Puzzle Palace. It's hunting and fishing now, remember?"

"Colonel Quinn?"

He didn't recognize the new voice but was suddenly aware that another party had been privy to their conversation.

"Yes, I'm Quinn. I didn't realize we had company." He didn't attempt to conceal the sarcasm in his voice.

"Sorry Caleb," said Henderson, "I should tell you I'm calling from the Situation Room of the White House and yes—there is company here—including your Commander-in-Chief."

Quinn sucked in a deep breath. *The Situation Room. The President. Sweet Mother . . . I've stepped in it now!* He swallowed hard.

"I understand, General."

Wiping at his brow with the back of his hand he felt the clamminess growing under his arm pits and guessed the other party was someone in his sixties. The man had a slight eastern accent.

"Colonel, as General Henderson stated, with our man Theodopolus hors de combat, we need someone we know on the scene. The information he had has got to be obtained—and quickly. You come highly recommended by Henderson. In addition, you've worked with us before so you're a known quantity, as they say. Will you help?"

Damn! They've got me between the proverbial rock and a hard place—and the rock is moving. And Karin—what do I do with her? Damn, damn, damn!

"Sorry, Mr., Mr.?"

Jocko's voice filled the earpiece. "That was the head of the Company." Quinn nodded to the wall. "Tell me," Henderson continued, "how far are you from Soudan, Minnesota?"

"Two miles, maybe three. Why?"

"That, Caleb, is where you're going. The information we need—and need tonight—comes from a site in Soudan, just above the town itself."

A light snapped on in Quinn's head.

"You're not talking about the University project at the bottom of the mine shaft—a search for quarks or neutrinos—something like that?"

31

"You have it right. You know the place?"

"It's been written up—friends have taken the tour."

"Colonel Quinn?"

Stunned at the familiar, youthful voice of the President, the smile faded from his face.

"Yes Sir!" His shoulders straightened.

"That project—like a lot of things we do—it's not what is seems. Security's been jeopardized and we need to get an immediate grasp on what's going on! A remote site, it's not likely we can get someone there in time to be of any use. Will you agree to help us?"

His shoulders sagging under the pressure of a direct presidential request, he nodded again to the wall and his Commander-in-Chief thirteen hundred miles away. "Yes Sir. I'm available." His voice was resigned.

"Thanks, Colonel Quinn. Perhaps we can meet after this is over— say down in Florida—I understand you're quite the fisherman—maybe we can swap lures?"

"Yes Sir. That would be just fine, Sir."

Henderson's voice filled the earpiece.

"Listen up Caleb. Give me twenty minutes at this end, then call this number back. You'll be given your orders and whatever background information we can provide. You have that?"

"Right, Jocko, but I should tell you. Maybe your problem isn't as great as you think."

"What do you mean?"

"Maybe somebody's already working a solution. This afternoon a convoy passed us, headed toward Tower-Soudan. It's made up of National Guard units from Range Cities with regular army advisers. A former enlisted marine under my command in Vietnam, now a captain in the Guard, gave me a little background and just a few minutes ago, I was told the objective *is* the State Park where the mine is located— that's the mission. Take the high ground, hold it from aggressors—a tank operation, in this case taking place in what might be the first blizzard of the season." He paused, looking at the heavy snow. "Oh, and by the way, this maneuver is loaded for bear. Live ammo all the way." He passed along Ernie's observations including the APC with unseen people and black boxes.

32

Henderson appeared to be talking to someone in the background. He returned to the line: "Make that ten minutes. And Caleb, you said there were regular army involved?"

"That's right. My best guess is twenty plus. They must be from the armored command at Fort Hood. The senior member is a guy named . . . Hartley. No, make that Hartlett, Charles, Colonel—you got that?"

Henderson paused, then responded. "Got it Caleb. Ten minutes then call back. Okay?"

"Ten minutes Jocko . . . ."

"And Caleb . . . get this job done and you'll have a basket of markers coming your way."

Quinn laughed. "Its' that bad uh Jocko?"

"You have that on high authority marine!"

*   *   *

Quinn waited in silence as the cafe owner placed steaming bowls on paper place mats. She moved the carafe of coffee, butter and a generous basket of warm, dark brown bread, smelling faintly of caraway seeds, from the serving tray to their table. The elderly couple, bundled up against the snow now blowing almost horizontally, paid their bill, leaving Quinn and Karin Maki sole customers in the pine-walled restaurant with gingham curtains. Sitting by the window, they could see an occasional car or truck move slowly down the main street through snow now almost six inches deep.

"What are we eating, Karin?" It sure smells good."

"Try *Karjalanpisti*. And if that's too big a mouthful, shorten it to *Karelian* roast. That's a Province in Finland. The bread's called *rekaleipa* and dunking is OK!"

Quinn dipped a spoon into the broth, retrieving a large chunk of meat from which fragrant sliced onions limply hung. "Can I call it Finnish Stew and not upset anybody?"

Laughing, Karin handed Quinn a chunk of bread. "You—Caleb Quinn—can call it whatever. And by the way—no liquor here so that Caleb Special—whatever—will have to wait."

After finishing the bowl and several slices of the bread, he pushed back from the table.

Karin smiled. "You seem more relaxed. Can you share anything with me?"

33

Quinn laughed. "Sure I can share. Because I really didn't learn a damn thing. Somebody's got a problem, but you know, I've got a theory . . . ."

Karin waited until the woman cleared the table and left a second carafe of coffee. She leaned forward, "Try me." Her blue eyes smiling.

Quinn lowered his elbows on the table.

"Conjecture; Somebody's got this thing under control already. I didn't get good vibes out on Highway 169, but I'm beginning to think Snowtop and his men didn't come from Fort Hood but are a reincarnation of the Green Berets. During my last two years in the Corps, I did some inter-service work with a new outfit called SOCOM—short for Special Operations Command out of Fort Bragg. The unit was airborne and included unconventional warfare specialists, behind-the-lines saboteurs, psychological warfare types and a civilian affairs battalion. If this project somehow relates to the Defense Department, it's possible somebody's dropped the communications ball. The Iran hostage rescue, Grenada, Panama, Desert Storm, Somalia— had some bad cases of missed signals between the services." He looked at his watch. "If I'm right, we'll find out in two minutes."

Karin kept her elbows on the table.

"And if? . . . "

Cracking a thin smile he pulled a twenty-dollar bill from his wallet. "We'll know that in two minutes as well."

$$* \qquad * \qquad *$$

Quinn hung up, his forehead throbbing against the cold, ice-rimmed window. His throat was dry. Wind driven fingers of snow eased underneath the cafe door. Water underfoot, a short time ago a puddle, was now a frozen sheet.

It couldn't happen! It only happens in the War College textbooks. But Jocko said it did—it was happening—and my information confirmed it. He turned, pushing open the door leading into the cafe.

Karin rose meeting him halfway between door and table.

"What . . . ?"

He shook his head, reaching for her down jacket. Helping her into the bulky garment, they left the change unclaimed and made their way to the Land Rover. Snow had already drifted up to the bottom of the vehicle. He unlocked Karin's side and trudged around the front to driver's side. Inside, he adjusted the heater to high, directing air vents onto the windscreen.

Karin broke the silence. "How much trouble are we in Caleb?"

Quinn stared across the bench seat, "Deep! I was wrong—dead wrong. Everything's gone wrong. The Russians are about to pull the rug from under us. Soudan is one of the most important research projects we've ever undertaken. Years of effort and the advantage of having the largest, fastest computers in the world, and now the Russians are shaking the fruit bush. The convoy we saw? It was headed for Soudan and a takeover of the mine."

"Russians, Caleb. What's going on?" Karin edged closer.

"The Defense Department knows absolutely nothing about weekend maneuvers in Northern Minnesota. The National Guard Command in St. Paul apparently processed orders coming from Washington calling for the maneuvers without knowing they were fraudulent. As Jocko said, when a surprise maneuver gets called everything clicks like clockwork. Who thinks to authenticate the paperwork?"

Karin pulled her legs up under her parka, the Land Rover was not noted for superior heating. "Who's up at Soudan? Those tanks? Your friend, the Captain in the Guard—I don't understand!"

Quinn pushed the gearshift into first. "There is a Colonel Hartlett assigned to armored command at Hood. The problem is that the *real* Hartlett's still down in Texas. The twenty or so other so called regular army types probably have an alter-ego stationed somewhere in the U.S. Remember what I just said about SOCOM?"

Karin nodded. "Green Berets, Special Forces types . . . .?"

"Russians are one big step ahead of us with SPETSNAZ, highly trained in all phases of conventional and unconventional warfare, assassination of military and civilian leaders included. As a service group, it dates back to the Russian Revolution."

"You're saying that, there *are* Russian soldiers at Soudan? How do you explain your friend in the Guard? He's not part of this takeover or whatever it's called?"

"It gets complicated, but follow this. Holmes gets his orders, as does the senior officer in charge of the National Guard troops in this area. They come through like all previous orders. Ernie knows the

35

regular army, working the Guard liaison office in Washington can authorize these maneuvers, so he assumes—as does everyone else—that everything is *kopacetic*, as the Poles say. The regular army personnel in this case are very professional, in command of each phase of the operation. They know exactly what they're doing and speak perfect English to boot. Ernie knows his next promotion is based on following orders. If this Colonel Hartlett tells him to place tanks at key points around the objective to be secured that's exactly what's going to happen. If he tells him no one is to get closer than 50 yards of his position without being challenged, Ernie knows that's the big test. Secure the objectives; hold it and rest assured that someone is going to try to penetrate the perimeter. He could care less what's happening *inside* of the ring—say, the project at the bottom of the mine shaft."

"It's scary, Caleb. How could the Russians get these men into the US—into our Army?"

"We've known about the modern SPETSNAZ for over twenty years. It's estimated to consist of over 30,000 elite troops handpicked for a variety of skills. We have documented proof that a number of their Olympic athletes are, in fact, SPETSNAZ. As such, they travel abroad under diplomatic immunity. Even their diplomatic staff, civilian and military, is made up of SPETSNAZ. Not only have they the advantage of travel to the country they're going to invade, they probably have spent a year or two at Krakowkyna."

"Krakowkyna—what's that?"

"In the Urals—a complete American City. Defectors have told us it's an unbelievable place. Cars are American and updated yearly, shops, clothes, food, strictly Made-in-The U.S.A.—from Levi's to Kentucky Fried Chicken. Depending upon your future assignment, you'll come out as American as you and I. Probably still operational even though the U.S.S.R., isn't."

Karin looked straight ahead, then turned to Quinn. "How does this figure with what's happened in Russia—the Cold War is supposed to be over?""

Quinn smiled. "Your thoughts coincide with the big hitters at the White House Karin . . . why would any government . . . particularly one in deep shit . . . do something like this? No answer—at least for now!"

"Where to now Caleb?"

"Soudan! Theodopolus lived in an apartment complex a short distance away from the mine. Then we track down another character that is with the project but for health reasons operates out of a rented home. We've got to pin down what the Russians have already done—what they intend to do in the next twelve hours. To complicate matters we've got keep a low profile. If Hartlett knows I'm still in the area he's going to immediately suspect something's going on."

Quinn turned the radio on then reached back behind his seat and fumbled for a few moments before retrieving a compact CB unit. He jammed the combination radio and police scanner into a metal track suspended under the dash to his left, made the quick connections and put the vehicle back into gear. Snow crunched under the wheels. Shifting gears, Quinn looked over at his silent partner.

"Legal or not, it's U-turn time again." Releasing the clutch, the Land Rover's thick-tracked tires gripped the foot-deep snow, crunching it under the wheels.

"One other thing Karin—effective with that last call, I'm back on active duty."

Karin was silent for a moment then reached out and put her hand on the back of Quinn's which rested on the gear shift knob. "Do I stop calling you Caleb?" she asked softly.

Quinn released his grip on the knob and held her warm hand tightly in his. He kept his eyes on the road ahead. "Its still Caleb—I'm going to need that touch of home." He shot a quick smile in her direction, squeezed the hand once and re-gripped the shift gear as the Land Rover started to slow down in the deepening snow.

*   *   *

The President stared at the map of Northern Minnesota on the large monitor. A series of rings circled Soudan, Minnesota. From his position at the head of the long conference table in the lower level Situation Room, he could see the concern on the faces of the men with him. His hand rested on a computer pad. Moving the device, a red dot shifted from St. Paul, to the center of the ring. "How many miles is that Andy?"

Boston Brahmin by blood line, Andrew Yoder, the thin, pale-faced Director of the Central Intelligence Agency responded; "240 miles over the road, 199 as the crow flies."

The President tugged at his throat. "I take it then—if troops were ferried up by helicopter we're talking about an hour plus flight?" He directed his question to no one in particular.

Chairman of the Joint Chiefs of Staff, Admiral Raymond Montgomery answered the question. "That's correct Mr. President. Airborne from Benning can be there in four hours but couldn't jump until the blizzard is over. Meteorologists say this one can stretch out fifteen, possibly twenty-four hours. In that same period of time, we can have two divisions—as many as twenty thousand combat troops ferried in from bases all over the U.S. They could be in the Minneapolis-St. Paul area, spread between the two airports ready to move out in minutes. If we deploy ground troops, they can be airlifted to the municipal airport at a place called Ely with a 4,000 foot paved strip that can accommodate C-130's—if the ground temperature has firmed up the subsoil. This would require a significant number of aircraft."

The President interjected with a raised hand. "Why so many aircraft?"

Montgomery flipped a sheet on the yellow pad before him.

"We have to move men from Ely to Tower-Soudan so each plane has to carry some form of light mobility—HUMVEE's, APC's, etc, in order to physically cover that distance of some 25 miles."

Frowning, the President shook his head. "Overkill! *Any* congregation of troops in that magnitude invites more questions than we can afford to answer. Besides, the weather isn't looking like an ally. Any other ideas?"

The Chairman directed his attention to the Commandant of the Marine Corps. "General Henderson, perhaps you'd explain your manpower situation."

Henderson placed his left hand over his right—the Annapolis ring glistened. "Mr. President, your Marines are the closest combat-ready battle group to the Tower-Soudan area."

"Marines? Your facilities are East and West Coast!" The President's eyebrows arched.

Fingers forming a bridge, the Commandant's two thumbs beat a slow tattoo against one another. "We have a designated NATO assignment in event of war in Europe—back up the Norwegian forces on the Kola Peninsula. To stay combat-ready, we train together

38

throughout the winter at a Guard Camp in Central Minnesota. The Norwegians send over troops and instructors, and training is conducted during the winter months under wartime conditions. A 600-man, combat-ready marine battalion is at Ripley along with four companies of Norsk mountain troops and their crack instructors. We have enough troop-carrying helicopters up there now to transfer two hundred men in one flight to Tower-Soudan in less than an hour. I've taken the initiative of contacting the Minnesota Air Guard's 133rd Tactical Airlift located on the field at Minneapolis. We've worked together on past operations."

Henderson tapped his copy of the twenty-page report. "Mr. President, your marines would appreciate the opportunity to take on the SPETSNAZ. Orders have been struck that would have the men at Ripley airborne on 15-minutes after we get a weather clearance!"

The President nodded in approval and his gaze swept the room. The other members of the Crisis Management Team shook their heads in agreement. He leaned forward. "General Henderson, it looks like your men get to do battle with the Russians—a first for the Corps—locking horns on our own soil." Turning to his left, he aimed his pencil in Yoder's direction.

"Andy, this is CIA all the way. I'm curious—how did we lose control of this supposedly secure project?"

Yoder looked across the table at the NSA whose face was impassive. Being on the defensive was a position the sixty-two year-old Spymaster was not used to—or comfortable with. He turned away from Hart and faced the President.

"I can't tell you *how* we lost it, Mr. President. I can only say this program, while extremely important, was deemed to be protected as well as it could considering the long-term nature of the study."

"That's an intriguing statement, Andy. Would you care to elaborate?"

Yoder glanced around the room, took a sip of water before beginning. "When Chen brought this Project to the Agency for funding ten years ago, he indicated it was an area that needed to be studied—that it was long, long term in nature. Further, the software hadn't even been conceived to run computers that didn't exist that the Project required. That decision was made by my predecessor—the best security

39

was to mask it with another study that *could* stand public scrutiny. Meanwhile we'd keep close tabs on its progress."

The President coughed lightly. "It would seem something happened to the tabs, Andy." It was said without rancor.

Yoder leaned forward. "Two things happened Mr. President. One, young Theodopolus teamed up with Doctor LaMont and suddenly the software side of the project began to accelerate. LaMont, apparently inspired by the infusion of energy and ideas from his new associate, came up with the idea of putting the CRAY Y-MP computers into entertainment—locking one to another to provide the capability a problem like this requires. What was supposed to have taken years, perhaps decades, to accomplish suddenly appeared to be close at hand. Breakthroughs came faster than we could digest them. It's Friday—but only this past Monday, LaMont sent word that a pilot-run on the Project was scheduled for tomorrow morning. Barring any last minute glitches—the first, full blown test will be punched out."

Pausing, Yoder glanced around the room at the silent members of the CMT. "This includes the first look at what LaMont and Theodopolus coined the 'LMRP.'" Yoder's gaze focused on the dark box, now heavily lined in the center of his yellow pad.

Hands on the cordovan folio; the President leaned forward on both elbows as far down the table as his body would allow. "You're telling us, Andy, that there's an actual time-frame attached to this computer run? We're not talking about hypothetical? This 20-page 'what-if' scenario actually has the potential for taking place?" The President fought to keep his voice under control.

Yoder's head bobbed.

"Doctor Chen, when he arrives, can elaborate, Mr. President. The Defense Analysis Institute, with help from my agency, keeps a steady flow of data coming from the other think-tanks and projects we know— or have reason to believe—are working the same problem."

Straightening up, the President retrieved the manuscript and quickly flipped through it. "Other countries know about this subject?" His brow furrowed.

Yoder nodded. "In the scientific community, Mr. President, its generally accepted knowledge, Russia, China, the Brits, France and East Germany, before it collapsed, have projects dedicated to the

40

subject. We've identified five sites in Russia alone that have a special focus in this area. The big difference between is that we own the most powerful computers."

"Until now." murmured the President.

"Until now," echoed Yoder, eyes downcast.

Yoder shuffled papers for a moment then added quietly. "There is a little more bad news I'm afraid."

The President's head snapped back from the map. All eyes focused on the CIA chief.

"Quinn's pass-along observations have been run through the ordnance file and his input about the black boxes leads us to believe that whoever is pulling this off has a serious number of Stinger Anti-Aircraft launchers in their possession . . . ."

The loudest sound came from General Benedetto and Commandant Henderson. The President, though not having a military background knew instinctively, this was bad news.

"How in . . . ?" The President stared at Yoder.

"We don't know where they came from Mr. President but the size described by Quinn is exactly that of a single Stinger firing tube with six rounds."

The Commandant whistled under his breath.

"That Mr. President is how we turned the Afghanistan War around. Stinger's in the hands of the mujahadeen brought the Russian Hind helicopters out of the sky like falling rain and kept their ground support aircraft flying too high to be of any use." Benedetto looked at Henderson as if silently communicating that they had a rocky road ahead.

"And Afghanistan Mr. President is where these Stingers may be from. We provided the mujahadeen, via Pakistan, 300 of these weapons with a matching compliment of rounds. We know that a number of them have filtered out despite the fact we are quietly buying them back whenever and wherever we can find them. The black budget cost is ten million dollars and rising."

Aware of the silent communication between Henderson and Benedetto, the President pondered the issue and finally asked the two men the question that was on his mind.

41

Benedetto answered. "This pushes our airborne penetration away from the mine site Mr. President. Until those units are taken out or somehow neutralized, whoever is up there has created a very, very potent defensive shield."

"This shield," asked the President, "how big is it?"

Henderson's hands spread out as if describing a bubble, "effectively Mr. President, I'd guess about five miles give or take the terrain considerations."

Military heads around the room nodded in concurrence.

"And what are our options Gentlemen?" asked the President quietly.

"Smart bombs from a distance Mr. President," responded Benedetto.

"Quinn's own resources to take them out as he finds them Mr. President," added Henderson, his eyes solemn.

*   *   *

"Prune juice mova'da pasta eh Pasquale?"

The heavyset CIA agent smiled. "Moved it big time. I can think positive thoughts again Kenny."

Sitting in the dark blue Ford sedan, the two inter-agency liaison officers compared the last of their notes.

"So its back in your hands uh?" The FBI agent shook his head. "And here I thought I had it made in shade for another promotion. Set the net for five errant scientists and move my ass up the big totem pole!"

"The Lord giveth and the CIA taketh Ken. That's the way of the world. But don't despair. I—we—need all the help we can get. These five brainy bozos appear to have been taken by someone—terrorists—who in hell knows? What we do know is that they all possess world class skills in the field of computer-aided design as it relates to topography. I now have been informed that there is a definite National Security issue—a major "*event*" is underway at a site somewhere in the Midwest. I get to shinny up the pole if I can find out where these four guys and a gal have gone and whether they tie to this Midwest situation."

Ken Waterman patted his mobile phone clipped to his belt. "I'll get back to headquarters and tell our folks to crank up to Ram speed. Call when you have something—I'll do the same." He opened the door and stepped out into the damp, Washington air. The traffic was beginning to pick up as people headed home early to prep for the Friday night party circuit.

Stretching from behind the wheel, the CIA agent extended his hand. "Great working with an old established business Ken."

*   *   *

Reconvening the marathon session after a break, the President looked around the table. Silver carafes bearing the presidential seal sat like small missiles ready for launch. "Are there any other ideas that need to be discussed?"

Hart raised his pencil in the air.

"You've got the floor, Tom," the President said, tugging at his loosened necktie.

The NSA pinched the skin of his lower lip. "We've spent the last two hours examining options using the best brains we've got and the guy recruited to do the job reached the same conclusion in a matter of minutes—Quinn said the Russians have got us by the balls!"

Hart, on loan from the prestigious Hudson Institute, continued: "The Russians apparently have had someone on the inside of the DANTE'S ORCHARD long enough to know when the computers were going to spill their guts so to speak, and now have the area secured. We can't bomb them out of the mineshaft; it's damn near impossible to do, and two, we won't because we'll kill the hostages. Pulling the plug puts our own people in jeopardy and besides, they have auxiliary power. All the Russians have to do is be there when the computers spit out the data, relay it to the Motherland—destroy the Project site—all four CRAY-computers and the one-of-a-kind software. They attempt to breakout and escape individually, die in a firefight or give up. Whatever, they've done their job. The irony of this whole mess is that we picked that location because it was remote, bombproof and easily defended . . . ."

The President leaned forward, tapping his pen on the table. "That sums up the situation, Tom, but the question I'm asking is, what in hell are our options? We've kicked massive manpower, firepower, air power—every damn form of power around. We can't seem to get to one solution cranked, much less having three or four to chew on."

43

Hart nodded toward the President. "I do have some ideas. Give this Quinn what he's asked for—an assessment of military resources available to him—direct access to this CIA analyst he's worked with before—Mumford—and we get the black-box types working on a crash program to block satellite communications in that area. The Colonel was as quick in that department as he was on the overall picture—when the SPETSNAZ get the data, it's going to be fed out via satellite just as quick as they can. I don't know how this ex-marine knows so much about satellite technology, but he's right on—first low-pass Russian orbiter showing up on the horizon after the data's collected and there goes the ball game—data's in the Kremlin—or some hole-in-the-ground they own."

General Henderson spoke, directing his gravelly comments to the National Security Adviser. "A point of interest. Colonel Quinn came up to speed on intelligence gathering when I had him review inquiry board findings regarding the Beirut bombing. His summary; too much intelligence in some respects, not enough of the right kind in others—and—lack of clearly defined command and control guidelines for dealing with what comes in over the transom."

Hart rubbed his temples. The President had long since learned it was a subconscious clue his NSA was drawing on his photographic memory. Eyes closed, silent words formed on his lips. The President felt uneasy. The bombing of the Marine Barracks and loss of 284 marines was not an event he cared to remember in great detail. Several of the dead were from his home state and, as state's attorney general, he'd attended the funerals.

Hart's eyes suddenly opened. He riveted his gaze on the four-star General. "There was a Quinn who died in that tragedy! General, is there a connection?"

The Commandant looked first to the President who averted his gaze, then to Hart. He nodded and spoke softly. "Second Lieutenant Kyle Quinn—Son—one month into his first duty assignment."

The room was silent.

Finally the President looked up, straightened his shoulders and looked directly at Henderson. He tapped the table in front of him with the slender pen. "As Commander-in-Chief, I accept ultimate command and control responsibility. What I *don't* accept however is poor

44

intelligence or intelligence that doesn't reach my desk. Now, let's get this meeting back on track. Hart! Do you have anything else?"

The NSA motioned to the world map. "A KH-11 satellite is being re-programmed to cover the site—as are low-pass ferrets. High and low satellites will be "*clicked-down*" for close-up coverage. We'll start to see infrared RECON photos within the hour from the 'boys in Suite 1000 at the Pentagon. Low altitude surveillance aircraft are being ferried to the air guard base at St. Paul—that's' as close as we can get with the weather as it is. Langley's been ordered to get its whiz-bang bird up and operational."

Swinging his gaze toward the Chairman, the President simply asked, "Ray?"

The admiral shook his head. "Same report, Mr. President. Naval Intelligence reports nothing unusual. No increased radio traffic, submarine movements, pattern shifts or increased fleet activity. As agreed, we've stepped up our security watch but are doing so as casually as the situation permits."

The President looked toward General Burton Smallzreid, Army Chief of Staff.

"Same report, Mr. President. All branches of Intelligence have been polled for unusual activity—nothing to report. Anything out of the ordinary will be funneled to Yoder. We're working with his people to determine how the SPETSNAZ infiltrated our system and pulled off the maneuver. You'll have a report by midnight."

The President remained silent for a moment, eyes still on the system that could put him in touch with the Commonwealth leadership. Head shaking, he looked around the room, puzzlement in his eyes. "We have Russian armed forces on our soil, in possession of one of our top secret operations and there is isn't so much as an increase in their heartbeat—a tell-tale blinking of the eye. We've agreed to a stand down in missile programming and the Country's like the proverbial mountain goat financially—precipice to precipice and back to piss again! Goddam, I just don't understand it," he said, slowly rising from his chair.

Turning, he left the Situation Room. Outside the door, he picked up his escort, a Navy Commander carrying a black leather brief case. Only the most attentive of observers could see the valise was handcuffed to the officer's wrist. Together, President and Bagman, carrying the codes for a pre-emptive or retaliatory strike against the Commonwealth, headed down empty halls beneath the White House. Walking slowly, favoring his left leg, he wondered how Colonel Quinn would react to a reading of the 20-page document-detailing Project V7162, otherwise called DANTE'S ORCHARD. *Would he be a believer?*

*　　　　　*　　　　　*

As Quinn laced up the felt-lined Sorrel's, Karin loaded the target revolver with .22 magnums. He worked his way back into his hunting parka, reversed from camouflage to duck-gray and placed a clip in the vintage Luger won in Saigon poker game.

"When you get there Karin, find out which unit Theodopolus lived in. According to Henderson, there should be communication equipment we're going to need. Ready?"

She nodded and slipped out of the cab, wrapping belt and holster around her hips, tucking them under the parka.

He looked at his watch."1821. Give it ten minutes then head back."

"What's that in real time?" she asked with a grin.

"Six twenty-one. Sorry."

As she disappeared in the snow, Quinn was privately relieved that she elected to stay.

Back in the Land Rover, heat on maximum, he spun through the radio dial, searching for a station with current weather. Nothing but static. Reversing the selection gave the same result. He ran a CB— static.

A heavy, metallic "Tap, Tap" on the driver-side window brought him up suddenly from his bent-over position and—inches from the long-barreled .357 Magnum—holstered the last time he saw it.

"What in Hell!"

The Deputy Sheriff, unsmiling, motioned Quinn out in the swirling snow.

"Slow—just move slow. Get outta'that rig and get spread-eagled. Come out with your hands up."

Quinn pushed the pistol under a sweater with a surreptitious movement of his right hand—lifting the door handle with his left.

46

Goddamn! Up to my ass in problems and this guy shows up again! What else can go wrong? Sliding out—mind racing for ideas, he left the truck running, coming out with arms raised.

"Why the gun, Deputy?" Change your mind about my illegal turn?"

The deputy backed slightly then steadied the revolver at Quinn's chest with both hands. Wind-driven snow eddied around the two men standing face-to-face in the darkness.

"No ticket! You're coming in for questioning—about the plane. Went back to look for tire marks and found some—in the cemetery—under the shed! We got a call into the FBI in case your buddy had something to do with snow from South America."

Time, damnit! Got to buy time. How long as Karin been gone? Will this guy pull the trigger?

"Be happy to talk with the sheriff—and the FBI. Put that weapon back in your holster. I'll follow you to your headquarters."

The Magnum jabbed into Quinn's sternum with force.

"Bullshit! You're getting cuffed. Turn around!"

Quinn remained stock-still, bare hands growing numb. The shadowy form he was hoping for materialized. *Time to distract our young friend . . . .*

He spoke slowly. "Forgetting something deputy?" He slowly lowered his arms.

"Damn you! Arms back in the air! I didn't forget jack shit!" The Magnum was inches from Quinn's nose. He crossed his arms. The .357 wavered, it's owner uncertain of what to do with the self-confident detainee.

"Forgot to read me my rights deputy. It's called the Miranda Rule—ever hear of it?"

"I know about Miranda damnit! You get that at headquarters. Get your hands up or I'll . . . ."

"Put your gun down deputy. NOW!" Karin's voice was crisp. Quinn was impressed by the command imperative. There was a moment's hesitation but the sound of a hammer cocking in his right ear, cold metal making contact with his exposed neck took on a meaning of its own. The deputy's revolver swung down in slow motion. Quinn retrieved the weapon, leveling it at the man's midsection.

"Karin, lift the cuffs out of his belt and put them on our friend here."

He waited for the "click. "You're a damn good officer. Hope to tell that to your Sheriff." The young deputy stared sullenly at Quinn.

"What do we do, Caleb? We can't leave him out here. And soldiers are already at the apartment."

Quinn looked at Karin. "Military—just great. What did you see?"

"Jeep parked in front—apartment's on the second floor—last unit on the northwest corner. I walked up to the second floor and pretended to knock on a door near the stairwell. A soldier was outside of his apartment—couldn't tell if he was Guard or not. How do you tell them apart?"

"You can't. Let's move deputy. You're going to tag along for the duration." Quinn nudged the man toward the rear of the Land Rover.

"It'll be cold, hopefully not for long."

Awkwardly, the deputy worked his way into the cargo section. Quinn locked the door.

In the truck, they shared the last cup of coffee.

Quinn looked at Karin's profile. "Hell of job back there. This marine was coming up short in the idea department. Thanks!"

Karin brushed a blond curl from her forehead; she didn't look toward Quinn. "Not too cool—Annie Oakley about wet her pants."

Laughing, Quinn dropped the vehicle into gear. "Let's go. We've got one fall-back location—better hope for some luck."

*   *   *

The house at the corner of Church and Center was dark. Bay windows over the garage faced north toward the mine. There were no lights or recent tracks—a relief to Quinn. The Land Rover idled. The window between cab and cargo section was open to allow heat back to their prisoner.

"Who lives here?" asked Karin.

"Man by name of LaMont, a scientist. Former head of the Project for the CIA but according to Henderson, had to step down for health reasons. He was a friend of the dead man—a mentor-student deal. Jocko said to make contact here if we couldn't get into the apartment."

"What if this LaMont isn't around," she asked.

"I'll go in and take a look."

"What are we looking for?" Karin wiped moisture from the window.

"Same as before—communications equipment. Theodopolus had a stash somewhere. Maybe LaMont was set up as a safehouse or maybe he knows where the Greek kid operated from. If this doesn't pan out, we execute an alternative plan."

"Which is?"

"Attempt to contact Holmes. Somehow we have to let the Guard know what's happening. SPETSNAZ won't hesitate to kill each and every one of them if their cover is blown. Second objective—find out what's going on in the mine. Henderson wouldn't—or couldn't tell me, except that sometime in the next twelve hours a bank of four computers—linked together—are doing something that must be damned important. Somebody in Russia must think so too, so much so that they've sent the SPETSNAZ in to intercept the data and relay it to Moscow—a suicide mission."

"How do they get the information out?" Karin flipped open the heat-vent at her feet. Outside, snow swirled around the edges of the house. Drifts were forming everywhere. Thick pine branches bent with the wind.

"Best guess is via satellite. Remember the rig next to the jeep?"

Karin nodded.

"Didn't think about it then but the antenna mounted on it could be the sending system."

Karin half-turned on the bench, pulling her legs up under her parka. "Can anybody stop the Russians from sending the data once they have it?"

"If—a damned big if—you know what they have to send with—*if* you know *when* they might send—and *if* you have any idea what they're sending to."

Pushing the door handle down, he shot a glance at the deputy and back to Karin. "Keep an eye on our guest. Looks too cold to be of much trouble."

Snow blustered into the cab. Swinging the belt with the Luger around his waist, he pulled the pistol free, flipped the safety and made his way toward the ground floor entrance. Flashlight in the other hand, he searched for a doorbell, found it and pressed the black button. He

49

waited and pressed again. He tried the door. Locked. Kicking through the snow, he yanked at the first garage door-then the second. Both locked. Peering in, he saw two covered snowmobiles and a van. Quinn played the light around looking for alarms but saw nothing. Returning to the door, he flashed the light up the dark stairwell. The door at the top was closed. Moving quickly, he broke the lowest pane of the door window with the butt of the Luger, wiping the glass free with a box-like motion of the barrel, reached in with a gloved hand and turned the lock. Working his way up the stair, a faint camphor-like odor caught his attention. Eye level with the base of the door, he stopped, looking for a glimmer of light. There was nothing. Crouching down, he laid the flashlight on the top step as he put his gloved left hand on the knob, the Luger held up in his right hand. He turned the knob slowly then with one swift motion, pushed the door open. He recognized the familiar smell of rubbing alcohol.

Hospital! It smells like a damned hospital!

A muffled sound came from across the room. Guessing the height of standing man, he flickered on the light. His aim was high. He lowered it pinioning a man in a wheel chair, eyes wide with fright. The white-stubble face stared into the yellow beam. An empty IV bottle, on a metal pole attached to the chair, swung to and fro. The tube appeared to Quinn to be still attached. A plaid blanket covered the man's legs but didn't hide bony, blue-veined ankles. Quinn rose, shifting the light away from the man's face.

"Doctor LaMont?" he asked quietly.

A feeble wave of the hand was the response.

He let the light play slowly around the room.

"Switch is behind you." The voice was tired and hoarse.

Flicking on the light, the man in the wheelchair remained frozen. *Another odor beside the sickly smell of medicine. Poor son-of-a-bitch must be incontinent.*

"Alone?" He asked quietly, lowering the pistol.

"I am now. But a friend is coming back—should've been here hours ago."

Quinn looked at the man but remained silent. His gaze swept the room. The walls were lined with books.

50

Papers covered two desks facing each other. There was a photo of two men standing side-by-side, arms over each other's shoulders. One wore a white knit sweater. The younger of the two was dark skinned. Quinn now knew what the man in the plane looked like in life—what the man in the wheel chair was dying of.

"Your friend, Doctor LaMont, wouldn't be Ted Theodopolus?"

LaMont nodded, a smile came to his face for the first time. He wiped his lips with the back of a hand, more bone than flesh.

"You know Teddy? You've seen him today?"

LaMont's eyes glimmered with life for the first time.

Quinn nodded wearily, sinking into the overstuffed chair.

        *   *   *

Looking around the cluttered room Quinn shook his head. *Karin's in this mess because I took her hunting, the deputy for doing his job, a dying scientist on our hands. Jocko old friend, can we cancel this marker?*

Karin removed her knit cap, shaking matted blonde curls free. "Told you this afternoon I wanted to be a nurse . . . ."

He turned to LaMont. "That OK with you?"

Fighting back tears, the gaunt man nodded. "I need some help but . . ." His eyes closed.

Quinn looked at Karin. "Doctor LaMont has AIDS."

LaMont's eyes blinked opened, but he nodded silently.

"The picture on the desk—LaMont, doctorate in geophysics with Theodopolus, his protege—was taken some time ago in front of a bar in Key West's Old Town—quarter mile from my conch house. It's the focal point for the gay community." He shifted his gaze to the scientist. "Do I have it right?"

LaMont nodded, hands trembling and resting on the handholds of his wheelchair.

"Jeez . . . AIDS . . ." The handcuffed deputy, sitting on the edge of a chair, stared at LaMont.

Quinn spun around, eyes glaring. "Karin, do what you can for the Doctor. I'm going to have a little talk with the deputy!"

"Tell me where to go Doctor," asked Karin.

LaMont nodded. "Straight ahead then the first left, Miss . . . ."

"Call me Karin." She pushed the wheel chair into the hallway.

51

Quinn walked toward the deputy who edged back in the chair. "Stand up!" he ordered.

The deputy hesitated, shrinking away. He yanked the man to his feet.

"Turn around!" he ordered brusquely.

The muscular deputy complied. Unlocking the cuffs, Quinn ordered the man to face him.

"Magnum's on the table. Leave if you want—but there's an order from the FBI notifying enforcement officers—where ever—to give me whatever help I ask for. This noon I was a retired colonel in the Marine Corps. One hour ago I was ordered back to active duty. There's a major job to tackle and I need men with "*sisu*." You look Finn—know what it means?"

"Guts-Sir." The deputy's pale blue eyes blinked.

"Right—guts like Big Frank's daughter. Now, you want out or are you willing to stay, keep your mouth shut and take orders?" Quinn backed away from the Deputy giving him access to the Magnum.

The young man swallowed hard. "I—I only have your word—I'm sticking my neck out . . . ."

Quinn anticipated the hesitancy of the young officer.

"Call 800-375-8700. Recognize it?"

The deputy nodded. "Law Enforcement Officers hotline to the FBI—for the State of Minnesota."

"Call it!" he ordered, pointing to the hand held unit clipped to the deputy's belt. "Tell them who you are and ask for confirmation. No other chit-chat."

"Rubbing his wrists, the deputy extended the antenna, leaving the Magnum untouched. The call was accepted on the first ring. The deputy asked the question as instructed, keeping his eyes on Quinn. The message both men heard was short and to the point. The deputy offered thanks and signed off, head shaking. "Don't know what in hell is going down Mister . . . I mean Colonel but I'm on your payroll—if you want me. Name's Saatela—Wally Saatela." Extending a beefy hand, "and I apologize for the heavy-handed shit I gave you."

Quinn motioned for the deputy to sit as he dropped onto the edge of the overstuffed chair. "Your background?"

Saatela chose to remain standing, assuming a parade rest stance, hands clasped behind his back, staring at the wall over Quinn's shoulder. "Twenty-eight—graduated from Ely High school—into Hibbing J.C.'s two-year law enforcement course—on to the army's guaranteed training program. Posted to Germany with the 17th Armored—MP duty eight months—Division Intelligence for the remainder of my tour. Joined St. Louis County Sheriff's Department. Been a deputy for two years. That's about it."

"Married?"

"Yes Sir—two years."

"Children?"

"One, and one on the way."

Quinn was silent, hoping he'd hear Saatela say he was a bachelor with no thoughts of marriage. "Weapons background—what can you handle?"

"Bolo Badge with the M16. Checked out with the bazooka and .50 caliber machine gun. Some experience with non-military explosives and of course the .375." He pointed to his weapon.

"Where does explosive training come from?" Quinn scratched at the wiry growth sprouting on his chin.

"Dad's a foreman with HibTac—blasting's his specialty. Taught me enough not to blow myself apart. Grandfather was a blaster too."

Quinn studied the young man. "Is there much in the way of explosives around here—say within a five-mile range?"

Saatela laughed. "Officially we know where all the registered dynamite is—and most of the unofficial stuff that finds its way out of the mines to blow boulders, take out beaver dams—converted to Dupont spinners when the food stamps run out."

"Sounds good. Let's put it to use." Quinn motioned to the Magnum. "I want you to line up whatever you can in useable explosives—don't take anything that's dated or looks questionable. We'll need primer cord, detonating caps and crimping kits. Also, we need more warm bodies. Got friends with Vietnam or Desert Storm backgrounds?"

Wally nodded. "A few. This weather—and deer season hours away is going to make tracking'em down a blue-balled bitch. Have to check every bar in the county to find some of these characters.

53

Diehards are already at hunting shacks getting buzzed—or playing poker."

Quinn frowned. "I want clear heads. Men that've been on both ends of a fire-fight."

"Service was all peacetime Colonel. Never shot anyone in my life."

Quinn studied the young deputy. "You've faced someone with a loaded gun haven't you?"

Saatela nodded with a smile. "Up here it happens once-twice a month—domestic abuse, mostly. Yeah!"

"Good! You'll do just fine. How much time do you need to find explosives and men?" He slipped on his parka waiting for Saatela's response.

"It'll take four—probably five hours."

Quinn reached into his parka pocket. "Take my rig—you may need it! Four-wheel shift is the red knob to the right of gear lever—with clutch in, pull straight back. Got that?" Saatela nodded, catching the keys on the fly.

"Get going! Do the best you can. I'll expect to see you no later then 0100 hours. And in the manpower hunt, Wally, if you can find anyone who knows about the mine itself—a big plus! The major problem is at the bottom of the shaft—ring of tanks at the top—something else to contend with."

Slipping the Magnum into the holster, Saatela paused. "My grandfather worked the mine for thirty years—knows every inch of it. When they shut down in the mid-sixties, he volunteered to go back and help turn it into a tourist attraction."

Quinn stood up, intrigued.

"Did he get involved in the work done for the University?"

Saatela was silent for a moment.

"Now that I think of it, yes. Worked for the contractor that hollowed out the main chamber—supervised drilling for the explosives."

"You said earlier he was a blaster too?" asked Quinn.

"Yes Sir—family tradition. Great-grandfather on my mother's side worked the mine as a black powder man before they had dynamite."

Quinn caught the look of obvious pride but he hesitated for a moment. Involving more than one member of a family bothered him but his need for knowledgeable men was immediate.

"Your Grandfather, Wally—he around here?"

Saatela nodded. "Two miles away."

"Think he'd talk to me about the mine? I need to learn as much as I can—as quickly as I can." He watched the young man's face as the smile faded. "Is there a problem?"

Saatela looked at the floor for a moment then back to Quinn.

"Grandpa and I don't communicate. I married a woman with Indian blood—and Catholic to boot! We haven't had ten words in the last two years."

Pursing his lips, Quinn asked, "Any objections if I contact him?"

The young deputy shook his head from side-to-side.

"His name?"

"Nick Huhta—dial 5841—don't need the prefix numbers. Will that be all sir?" Saatela zipped up his blue jacket, pulling on his heavy gloves.

"One last thing—bed sheets—white variety only—we need a bunch of them. Tell whomever you get them from, they'll be reimbursed later."

Saatela saluted smartly, a smile returning. "Men, dynamite and white bed sheets—Sir!"

*   *   *

Quinn moved through the cramped study searching for the location of communications equipment used by the dead scientist. *If Theodopolus and LaMont have radio gear, they've hidden it well.*

His head grazed the replica of the ancient bird of prey. Grasping the balsa model, the string gave slightly. The television monitor suddenly came to life. Tugging again, the screen went dead. Another tug and the set re-activated. He smiled. *That's part of the package.*

Moving quickly, he headed for the globe over the PC's. Tugging produced no results. He stared at the smooth surface. It was unlike any globe he'd seen-devoid of continents or oceans—just black and white sworls—a giant thumb print. Moving to the last globe hanging in an unlit corner, the basketball-sized orb was given the tug test. The effort was rewarded with a soft "*click,*" then a slight hum. Snapping on the flashlight, he swept it over the book-lined shelves. Nothing seemed out of the ordinary. He tugged at the globe again, alert for sound and motion. Peripheral vision caught the movement. *I'll be damned. Angled shelf in the library stacks, recent magazine showing. Back issues underneath.*

A small section of the bookcase, waist level high, was open exposing a compact radio console with a headset. Dropping to kneeling position, he eyed the panel. *Expensive! Must have been made in Company labs, tied to satellites somehow.*

At that moment, Karin wheeled Doctor LaMont into the room. Quinn rose, facing the geophysicist now dressed in khakis and a sweater. His gaze darted from Quinn to the radio and then the glowing monitor. The man's knuckles tightened on the arms of the wheelchair, face growing dark.

Karin, sensing the rising tension, broke the silence.

"What's happening? The TV, it's on but not showing anything."

Quinn reached up pulling the globe twice in rapid succession.

"Jocko said there would be communications and surveillance capability. I found a radio and the means to activate what must be a remote controlled camera while you helped *GRANNY SMITH*." He watched LaMont's face for an involuntary signal of admission. It came in the form of an almost imperceptible blink of the eyes.

"GRANNY SMITH? What are you talking about?" Karin's gaze shifted from one man to the other.

Quinn pocketed the flashlight. "Karin, meet the senior member of the CIA's two-man team inside DANTE'S ORCHARD. The man in the plane was Ted Theodopolus, LaMont's protege . . . code named . . . JONATHAN . . . as in the apple."

LaMont, exerting himself, suddenly leaned forward. Karin lunged to keep him from getting out of the wheelchair.

"Was? Was? What happened to Teddy?" Rasping, fighting for breath, LaMont's face was one of uncontrolled anguish, his eyes flared open.

"Easy—easy, Doctor." Karin gently restrained the scientist.

Approaching LaMont, Quinn pulled the leather chair around. He was silent for a moment, gathering his thoughts. When he spoke, the words came slowly; "Doctor—your friend—Ted Theodopolus died this afternoon. *Termination by extreme prejudice'* as your people say—shot in the head at close range while he waited for me . . . probably late afternoon. Shortly after, a thermite device destroyed the plane and consumed his body. Enough remained for the local police to know someone had been in the aircraft." He clasped his hands, gaze solidly on LaMont's drawn, tortured face.

"Under the circumstances I'm coming to understand, I can appreciate your grief. Teddy Theodopolus was more than a fellow agent and gifted student . . . ?"

The scientist nodded. "You understand correctly . . . we were life partners." Words barely audible, LaMont's eyes closed.

Karin began massaging the neck and shoulders of the grief-stricken man. She held back the tears.

Quinn leaned forward. "I can't ease the pain over your friend's death but bear with me. Six hours ago, I was minding my own business. Now, I'm up to my armpits in something I don't fully understand, re-activated to military duty—and my newly assigned code name is GATEKEEPER . . . ."

LaMont's eyes blinked open. He wiped at them with the back of the woolen sweater. "Would that be GATEKEEPER as in one word or two?" he asked.

Quinn looked at the ailing man, then slowly spelled out the word letter-by-letter then added, "One word, no hyphen."

LaMont looked up at the ceiling momentarily then at Quinn as he expelled a sigh of relief. "Welcome to the ORCHARD'S topside outpost."

Leaning back, Quinn studied the researcher in silence. "Time, Doctor LaMont is *not* on our side. Your friend and associate is dead and the project appears to be under complete Russian control—inside the ORCHARD—and out! At the moment, SPETSNAZ control over 200 National Guard troops—those troops—plus twelve tanks and seven APC's block any possible access to the lab. This storm is one of the biggest ever seen in the upper Midwest and its got a choke hold on any military support being sent up here."

LaMont's head moved from side-to-side as if to deny what he was hearing.

"My assignment is to stop the SPETSNAZ. To do that, I need to know everything you and Theodopolus were involved in—find our remaining assets—identify the liabilities. Our Commander-in-Chief is sweating bullets—wants Moscow on the Hot Line now! His advisers are telling him hold off—until I call in the damage report and give the Crisis Management Team a clearer picture of what went wrong and some idea about what's currently happening."

Quinn's gaze swept over the room, head shaking at the computer array, terminals, globes and books. "It's time for a stiff shot of Crow's Maxim . . . ."

Karin and LaMont looked at Quinn's half-smiling face.

"Do not think what you want to think until you know what it is you need to know . . . from the book by Mumford . . . ."

He looked at the geophysicist. "What is DANTE'S ORCHARD all about? What have we got that's so important the Russians would risk nuclear retaliation to get their hands on?"

LaMont fixed Quinn with a hawk-like stare. His left hand gripped Karin's wrist, tugging her to his side. "You've given me the correct code name and qualifying response Colonel Quinn. I'll buy that *you're* the Company's designated representative. But what about her? Who is she? Why is she here?"

Despite his weakened condition, LaMont kept a tight grip on Karin's wrist.

Quinn leaned back, a faint smile creasing his face.

"Karin Maki is a friend who had the great misfortune to accept an invitation for an afternoon hunting trip. So far she's out-gunned me in the field—pulled my biscuits out of hot coals once already today and is now your private nurse. She's in this with me. My clearance is UMBER-A and I'll take responsibility for what she's exposed to. You want to check me out—do it but bear in mind it'll take time—the Russians may have this system tapped and as I said suggested moments ago, time is critical."

LaMont's shoulder's seemed to sag visibly as he released his grip on Karin's arm.

Quinn clasped his hands, leaning forward again. "Again . . . what in hell is this Project all about. What's in the ORCHARD under our feet the Russian military has come to harvest?"

LaMont's left hand rose slightly. "I apologize Colonel. Between my physical condition—what's happened in the last ten hours. I'm having trouble sorting things out." He looked up at Karin. "Please sit down next to the Colonel, I'll begin by . . . ."

"Caleb! Look! The TV!" Karin pointed over Quinn's shoulder. LaMont's head jerked in the same direction.

A dark shadow filled screen.

"The focus controller Colonel . . . between the PC's!" Doctor LaMont motioned with his gnarled hand.

In a swift move, Quinn had the joystick in one hand, volume control in the other. Quickly, he twisted the knob clockwise. The screen blurred. He reversed the motion. Hands, followed by a pair of dark almond eyes filled the screen.

"Damn," murmured LaMont. "That's Vidall. He's discovered the surveillance camera imbedded in the ceiling. It's thirty feet off the floor—he must be on the cherry picker!"

Suddenly the screen went blank.

"Why can't we get sound Doctor?" Quinn violently twisted the volume control up to the stop, static the only response.

"It's been disabled."

He spun around to face LaMont. "Can they trace the camera back here—is that the only system?" The words came rapid fire, his gaze riveted on LaMont's blinking eyes.

"The only system—technicians down there can trace it now that it's been located. When they track the wiring, it will lead to a mini-microwave transmitter which points down hill to the same sender-receiver system mounted on the corner of this house. Line-of-sight, they'll figure it out pretty quick! It was installed during a period when no one else was allowed in the main computer room."

Quinn rubbed his forehead, looked at Karin and then back to LaMont.

"Look Caleb, Doctor LaMont . . . look at the screen now!"

Karin pointed over Quinn's shoulder. They stared at the monitor that had suddenly come back to life.

Quinn twisted the knob bringing the floor of the cavernous computer room into sharp focus then turned the traverse control. There was no response. The camera remained immobile.

59

"Damnit . . . how stupid! . . ." Quinn's hand left the joystick as if he had grasped the wrong end of a hot poker.

"What's wrong?" Karin's eyes shifted from the screen quickly to Quinn.

"Now they know the camera system *is* operational and being manipulated. When I used the zoom they could tell somebody—somewhere was using it. Traverse capability's been frozen but for some reason they haven't cut the signal output."

Quinn turned to LaMont.

"How long does it take to get here from the mine shaft entrance?"

LaMont rubbed at his skin, shaved smooth by Karin.

"Teddy could get down there on his bike in fifteen minutes. Going back it was thirty minutes—uphill. In this weather . . . ."

Grim-faced, Quinn looked at his right hand, his thumb bounced off each finger in rapid succession ticking off minutes and options and then back to LaMont as if appraising the man. "Unless I'm guessing wrong, we're going to have visitors and damned quick."

LaMont's face went slack. "How do you figure Colonel?"

"Easy. How many members are there involved in the project?"

"Forty," he responded.

"How many got nailed when the takeover occurred?"

"Thirty-eight," the instant reply.

"Taking Theodopolus out of the picture leaves how many unaccounted for?"

LaMont nodded weakly, "I get the point."

"Do the people in the Project know where you live?" snapped Quinn.

"Yes. In the old days . . . before this . . ." LaMont tapped at his chest, "I had parties here. Gave them up three years ago. Today my involvement is very low key."

"Another question. Is it common knowledge you have AIDS, that you and Teddy were . . . close friends?"

LaMont shook his head. "I don't think so. It was politic to keep my condition out of the limelight. When it was diagnosed, the Company and I, Teddy concurring, passed it off as multiple sclerosis. Left my position as on-site coordinator for the project, setting up shop here. Teddy worked out of both locations. He rented an apartment a half mile away but it was mainly to keep people from knowing he spent most of his nights here. Just another cover."

Quinn rubbed his chin, digested the information and fired another question at the geophysicist; "How critical are you to the Project? Can the people down there operate without you?"

LaMont, cocked his head at a quizzical angle and was momentarily silent. "Hadn't given that any thought. Teddy was totally up to speed with our end of things. If there was a glitch, he would do the hands-on work down there. I'd work on my computer up here and feed data to him. That's how we worked—me up here, Teddy down there. Nights and weekends, we'd work together to program and polish. With Teddy gone, yes, it's possible I could be critical to the mass."

Quinn glanced at the twin computers. "These PC's tied into those underground?"

The doctor nodded. "They were until eight this morning. While Teddy was fixing breakfast, I punched up my unit to hook into the CRAY's and was denied access. He checked his—same thing. I can guess what you're thinking—can we sabotage the program?"

Quinn nodded emphatically.

LaMont answered his own questions. "Theoretically, the answer is yes. *If* I could access the system. At the very least, I could throw up some barriers the others would have to hurdle. We did that from time-to-time to test the software team, Bochert's Burro's on the day shift, Wolf Pack at night. In most cases, the whiz kids could find the glitch in minutes. Once in a while, we could stump them for an hour or two—never more than that."

"So the answer is Doctor—*if* a problem comes up—or can be created—you *might* be indispensable?"

LaMont nodded. "I should explain Teddy made a major breakthrough about six months ago, which involved the problem of linking our CAD program to the TBM."

"TBM? Explain Doctor—CAD I'm familiar with."

"Time Base Matrix. Once events start to occur, geophysically, the LMRP has to be in perfect step. If it isn't, one set of data is worthless."

Quinn turned toward the window, pointing Karin in the direction of the wall switch. She turned off the light and the room was dark except for the pale light of the monitor. Quinn shook his head. "Let's

61

back up a moment. Time Base Matrix I can roughly understand, Doctor. What's this LMRP you mentioned? Seems to be a key element. What's it all about?"

"Action on the monitor, Caleb." Karin called their attention back to the screen. A tall, thickset soldier in winter field dress stood in the center, one hand resting on the holster on his right hip. Field cap off, his head was white-haired, close-cropped.

"Snowtop," Quinn murmured to Karin.

The man with the flat accent gazed upwards into the camera. Suddenly, without adjustment, the static disappeared—his monotone voice came out of the set.

"Doctor LaMont. While I suspect you can communicate with me, no response is expected. I've taken the liberty of sending a jeep to your residence. With the Project so close to your heart, I wouldn't want to deprive you of the opportunity of being here when the two programs are joined—courtesy of these wondrous CRAY Computers." The man smiled, a mocking look on his face. "My men are aware of your *fragile* condition but please don't attempt anything heroic. There are people down here dependent upon your cooperation." He motioned to someone out of camera range. Two soldiers appeared supporting a third man, a civilian, head hung over his chest.

"Damn! That's Bochert. What in hell have they . . . ?" LaMont's voice faded in silent shock.

Grasping the man's hair, the white-haired officer snapped the head back to face the camera. The team leader's face was swollen and bruised, eyes puffed shut. A thin, dark stain ran from his mouth down to his neck and disappeared in the "V" of his woolen shirt. "Ten minutes, Doctor LaMont. Bring whatever medical supplies you need with you. And by the way, I've taken out extra insurance as I believe you say here—observe . . . ."

The camera swung into a corner where five people stood huddled—four men, one woman. The camera closed in on their faces. A voice off camera ordered them to look at the lens.

LaMont groaned and flopped his hand in a futile gesture. He looked up at Quinn who was pulling on his parka. "Who was that man?" he asked hoarsely.

"Goes by the name of Colonel Hartlett. However, the *real* Hartlett is still at Fort Hood, Texas. Your people are trying to figure which SPETSNAZ commander is using his name—and how he and his men got into this country. Who were those people—you seemed disturbed when you saw them?"

"Whoever the Colonel is, he or the people he works for have got their hands on the best and the brightest in creating high-performance numeric computation. Theodopolus's peers. In one room you have the world's most talented designers of integrated symbolic computing software—linear, non-linear, 2-D, 3-D, Neural Networks, Spline Analysis, morphological operations, RISC interface . . . ."

"Whoa LaMont! You lost me on the first volley. In short, those five people can do what you and your friend have been doing?" Quinn stared down at the Doctor who shook his head in the affirmative.

Quinn turned quickly to Karin. He spoke rapidly. "Doctor LaMont isn't going anywhere. Five brains or not, they must need the good doctor. While I prep him, you bring out the dextrose hook-up and any other paraphernalia used to sustain him. We've got to convince our SPETSNAZ commander, as vital as the Doctor may be to the project, it's imperative he remain here. 'Medical supplies' as they put it, had better be out here in front. We've only got minutes to set this stage."

Quinn dropped to his knees in front of the doctor. "What I'm going to ask you to do is distasteful—down right repugnant but our options seem to be slim and none! Everything you need to survive is here and the communications line is all I've got linking me to the ORCHARD. I'm betting that when Hartlett finds out you have AIDS and gets a true picture of your condition, he'll re-think having you down in the mine. Do you understand what I'm leading up to?"

LaMont stared quietly at Quinn. A moment passed. Finally, he spoke, his voice resigned. "Play the role of a leper?"

Quinn nodded sadly, "One very near death Doctor LaMont."

*   *   *

The Situation Room had been transformed with additional FAX machines, copiers and monitors. Multi-colored phones proliferated over the conference table. Half-eaten sandwiches and coffee cups littered the rectangular room. The stale stink of smoke lingered. Director of the CIA, Secretary of Defense and State converged toward the table as the

President re-entered. Hart approached the President with a small, white-haired Asian in tow.

"Mr. President, Doctor Matt Chen of the Defense Analysis Institute."

The President extended his hand. Matt Chen looked up at the tall man, adjusted the spectacles as if to make sure of his identity and carefully deposited an ancient leather valise at his side.

"A pleasure seeing you again Matt. Given the choice, I'd rather hear you play Mozart on the Steinway upstairs after a pleasant dinner."

A wide smile of remembrance over their last meeting lightened the little man's face. Chen grasped the President's hand warmly. The 70-plus geophysicist traced his heritage back to a great grandfather who helped build America's first transcontinental railroad and beyond. Few knew that "Matt" didn't stand for "Matthew" of biblical fame but rather "Matteo"–an Italian Jesuit priest who helped open the doors of China to Catholicism in the early 1600's. The head of the Defense Analysis Agency reverted to a slight style bow after releasing the chief executive's hand.

Occasionally, a fax machine would operate in the far corner. Yoder, sitting the closest to the transmit-receiver units, received the incoming traffic, reviewed it, folded it and send it down the table initialed for the correct recipient.

Glancing to the far end of the room, the President checked the bank of clocks. 9:01 PM Eastern Time. The Seth-Thomas to the far right, Moscow time, read 4:01 AM. Conversation had come to a halt. They waited for the President's signal. He nodded to Yoder who carried a device the size of garage door opener. He pressed a button on the unit. Instantly, a low level of sound permeated the room, not loud enough to disrupt conversation but powerful enough to provide a sonic shield against intrusive ears.

Reaching for the leather folder, the President flipped open the cover exposing the light-blue notepaper. The penned question mark had become fat with repeated tracings. "Satisfy my curiosity Doctor Chen. Who is 'AWG?'

As if expecting the question, Chen reached down and retrieved a manila folder from his valise.

"This, Mr. President, will tell you who he *was* . . . ."

The President's eyes scanned the print. "Amadeus W. Grabau, S.D." he announced to no one in particular. He brought the sheet closer. "Fellow of the Geological Society of America, Paleontological Society, New York Academy of Sciences, Academy Sinica." He paused, "American I presume." Chen nodded.

The President eased the sheet away. "Research Professor of Geology, China Foundation." He read the next line in silence, then looked up at Chen who nodded in anticipation, a slight smile on lips he tapped with his index finger. "Chief Paleontologist, Geological Survey of China?" The President's voice rose involuntarily. A murmur floated above the white noise. Doctor Chen's head bobbed and he fished for the paper beneath the folder, passing it to the President.

"*The Rhythm of the Ages . . . Earth History in the Light of the Pulsation and Polar Control Theories . . . .*" The President stopped. He looked at the leather folder, the second sheet and back to Chen's impassive face.

"The bottom of the page, Mr. President, please."

"Published by Henri Vetch—Peking—1940?" The President looked at the Professor who simply nodded again.

"So this Grabau—S.D.—whatever that means—is responsible for this twenty-page doomsday scenario that says we're going to be destroyed by a polar shift, that the world as we know it will be devastated, and—when it starts it could be over in a matter of hours?"

Doctor Chen said nothing. The Situation Room was silent save for the almost imperceptible hum. He pushed the thick glasses up the bridge of his nose. "Grabau was a theorist. Not the first to propose that the world has experienced polar shifts with calamitous results—most certainly not the last. His major contribution was to provide researchers who followed, a footpath of sorts, to better understand *how* and *why* such events took place—and clues to what conditions could precipitate it happening again. He didn't, in 1940, anticipate such situation as the *'greenhouse effect'* or nuclear testing. Grabau avoided the pitfall of predicting *when* it would occur again. Remember—the best computer in the world at that time was still the abacus. But young Theodopolus used Grabau's building block, *Rhythm of the Ages* and high speed computers to create a multi-disciplinary program which is on the verge of being operational . . . ."

George Weller, jowly, well-liked Secretary of State, directly across from Chen, rubbed a mottled right hand over his left. "I presume Doctor, DANTE'S ORCHARD has everything to do with the *'when'* factor?"

"Precisely Mr. Secretary. By way of explanation . . ." He pointed down the table toward Admiral Montgomery." The Navy's Office of Oceanography is responsible for providing world wide tidal data. Formulas for determining ocean and earth tides earth have been with us for over a hundred years. Until the late fifties however, oceanographers used a wondrous, brass cog and wheel device, big as a room, to grind out tidal charts for a given location. With the advent of computers, the process was reduced from days, to hours and minutes. Now, with CRAY, we talk in terms of seconds."

Chen paused, removed a handkerchief wadded in a pocket of his vest, dabbed his glistening forehead, slipped off his glasses and dried the bridge of his nose. "The linkage," he continued, "between being able to predict tidal effect in say, Tahiti on a given day, hour and minute and what is happening in DANTE'S ORCHARD is crucial. Gravitational forces, caused primarily by the moon, create Earth Tides. Not only does the moon, and to a lesser extent, all other planets in our system pull at the oceans, causing them to rise and fall, but the *very* mantle of the earth is subject to these forces. These forces are so strong that the earth's outer shell flexes twenty inches in the span of one lunar day—twenty-four hours and fifty minutes. In calculating these omniscient forces, thirty-six constituent factors must be considered, taking the gravitational input of each planet—Jupiter, Mars, Saturn, and so on, and assign them a value as part of the vector formula."

Hart, to the left of Weller, raised his pencil. "Doctor, tidal forces I think I understand, being a blue water sailor of sorts, but how do Earth Tides relate to this Doomsday scenario?"

"Directly Mr. Hart. The lynchpins are the Earth Tides and Plate Tectonics. Plate Tectonic researchers, like Doctor LaMont, believe that Earth Tides will be the primary triggering mechanism to cause a polar shift." Chen looked at the President and blinked. "And also, my apologies. 'S.D.'—it is an old fashioned term meaning, *'Scientaie Doctor.'*"

66

He pushed the glàsses up and went on; "A geophysicist, not a geologist like LaMont, I *do* however, subscribe wholeheartedly to the theory of Plate Tectonics. Continents rest on massive moveable plates which have moved in the past with catastrophic results—and most certainly will do so again in the future. Witness the earthquake in Armenia several years ago—tremors here on the East Coast—volcanic action in the Philippines, more activity in California . . . ."

"The speed with which such event takes place Doctor—can it truly happen in a matter of hours?" queried the President.

"Ah, yes! Paleontologists can make a strong case for the fact it has happened quickly—the discovery of pre-historic animals, mainly in Russia, preserved in ice with undigested greens still in their stomachs is evidence of this. Between history and archeology pre-dating biblical times, there is much to suggest that polar shifts have occurred before— and with instantaneous and world-wide repercussions."

"China?" The President asked as if the thought had just occurred, "have we been looking in the wrong direction for the source of company in DANTE'S ORCHARD?"

"Ah No! Mr. President, but perhaps it would be appropriate for Mr. Yoder to explain."

Yoder acknowledged Chen with a nod. "Files on this subject, pre-date 1940. Sketchy at best and then—in the hands of J. Edgar Hoover's agency. Because of the global implications and the soon-to-be-formed, OSS—Operation of Strategic Services-the subject matter was transferred to our predecessor organization. With World War II coming to a boil, Grabau's work took on a certain timely significance. The names referred to are: Ela Kravchenko, a man known as Son-Wong and a Professor Kleinhauser—all long deceased. Kravchenko, a White Russian, we believe worked for the NKVD, precursor to the present KGB. Chiang Kai-shek's secret service may have leaned on Son-Wong and there's some reason to believe Kleinhauser *might* have been pressured into advancing data to Hitler's *Abwehr* when he returned to Germany in 1941. Result, Russia, China and Germany most likely came into possession of the data the same time we did. The three most powerful countries in the world at the time, started off on even terms. If Hitler had access to the data, we don't know if his people paid much, if any, attention to it. Our file was dormant for the next three years. It was

67

re-activated in 1944 on the advice of Robert Oppenheimer who was concerned that the Atomic Bomb could somehow trigger a polar shift."

Yoder paused, aware his last statement had caused a stir. He glanced from President to Chairman of the Joint Chiefs of Staff. "A top secret study, still classified and prepared in the early sixties by Chen's predecessors, discussed in depth, use of nuclear weapons in the polar regions to cause de-stabilization in certain portions of the world. As an aside, I might add that we're looking closely at the results of our first space-mapping program. As you know the high-resolution results are for *'military-eyes-only'* and the preliminary assessment is that we're seeing evidence of more ground faults than we heretofore have known about and . . . can see that polar destabilization has advanced more rapidly than the scientists have thought possible. As you've probably heard—we've recently picked up on the fact that an ice floe twice the size of Delaware has broken off Antarctica and may be a harbinger of even greater pieces yet to come. God knows how this will affect situations like shipping and regional meteorological conditions . . . ."

Dr. Chen raised his hand and the President nodded.

"To let all of you know that studies such as the University of Minnesota and Cambridge are conducting are not isolated situations, China and Italy have long cooperated on a similar study at Gran Sasso. The stated purpose at *their* underground facility is to prove—or *disprove*—the existence of "WIMPS"—restated as *'Weakly Interactive Massive Particles.'* It's not impossible that there is some linkage between the current project and Grabau's earlier efforts of the late 1930's . . . ."

Disturbed by what he was hearing, the President turned to Chen: "If I read my file correctly—and I've read it five times—we're talking about something completely out of our control. History runs in cycles, the universe in rhythms large and small, and we could get blown off the globe in a snap of the finger! Am I understanding this correctly, Doctor?"

Chen's head pumped up and down. "Ah, that is so! There are rhythms, to be sure. But other factors must be taken into account. I've mentioned the *greenhouse effect*, the saturation of the atmosphere with chlorofluorocarbons—rapid global warm-up. This by itself could trigger a polar shift. We now have still another problem—the detritus of war in the Gulf and the massive injection of solids and sulfuric acids from Mount Pinatubo and host of smaller volcanoes since."

The President and the men in uniform leaned forward almost as one.

"How does Desert Storm figure in this scenario Dr. Chen?"

Chen looked around the room as if assessing the risk of offending anyone present. "Observations Indicate that the heavy bombing in the weeks prior to the actual 'war' churned up massive amounts of fine residue which were picked up by the Shamal Winds which in turn were coated with the oil from the Kuwait fires. Literally thousands of tons of this material are in the world jet streams, suspended—it seems forever—a meteorologist's version of the Flying Dutchman. We have no way to quantify the impact it is or may have on a world scale. This is an event unparalleled in the annals of man—how it links to the work of LaMont and Theodopolus—the data is just now coming in . . . ."

"Doctor Chen. Having covered this subject from many angles, tell me what you can about China's position on this issue." The president leaned back, indicating with a flick of his hand that the researcher once more had the floor.

Chen put his pen down, glanced at his notes and then responded: "China has to be concerned with issues having to do with plate tectonics Mr. President. It sits astride numerous earthquake faults. A country with a long history, it's kept good records on the subject. An earthquake which occurred in 1556 killed over 800,000 people–the one that happened in 1978 killed an estimated 500,000 Chinese."

A soft murmur from around the table greeted Chen's statistics. The president's eyes widened.

Chen continued: "And, a creative Chinese by the name of Chang Heng invented the world's first seismograph in 132 AD."

The president leaned forward. "Did it work?"

Chen laughed. "Quite well Mr. President. But the Chinese have more than lives to be concerned about when it comes to disasters such as earthquakes, floods and droughts however . . . ."

"And those concerns might be . . ." the president responded.

"*Mandate of Heaven* Mr. President. Dynasties have fallen when natural disasters of a large magnitude took place. The people would rise up, rebel and push those in power off the throne.

The president laughed quietly and looked around the room over the top of his glasses. "Nothing new there eh fellows?"

His question was answered by several "*amens*" and much nodding of heads.

Nodding wearily, the President raised his hand, a silent plea for respite. "A five minute break Gentlemen. I need some fresh air and a few moments to digest what I've just heard."

*   *   *

Resplendent in his formal attire, the young Navy lieutenant, kept his hand lightly on the First Lady's elbow, guiding her away from the black-tie crowd of literati around the edges of the parquet floor. Inwardly, she was seething. Strong willed and self-possessed in private, she needed the calming influence of her husband in public. Leaving the ballroom, they passed a man in a double-breasted evening jacket and his raven-haired companion who raised glasses to the departing hostess. She knew them as Mordecai Baratz, Washington correspondent for Tel Aviv's Davar, and his 'wife,' Laniya, a visiting professor at Georgetown University. In their native country they were known as Brigadier General Baratz and Captain Cohen—married—and assigned to the Bureau for Scientific Affairs, otherwise known in Hebrew by its acronym, LEKEM, an office of the Israeli Defense Ministry—an appendage of Mossad, that Country's intelligence agency.

Baratz quietly discussed the early departure of the President with Cohen. The First Lady, leaving with a promise to return, piqued his curiosity. Splitting up, Baratz moved across the room, searching for a White House staffer, male or female, susceptible to his charms. Cohen, tawny skin set off by a black sheath dress, glided toward an Air Force major in his trim mess jacket positioned near the door for ready conversation for unattached women. Cohen was confident she could hold the man's attention as long is it took for the First Lady to return. Fingering the gold Star of David, warmed by mounded breasts, Laniya flashed an open-mouthed smile at the officer as she took up her station.

*   *   *

Dismayed at ending the conversation on a jarring note, the President, chagrined, stared at the phone. Upon telling his wife that the time of his return was unknown, she had hung up.

70

At the door to the conference room, the President stopped. "Have we had any word from our man, Quinn?" he asked of the NSA.

"Nothing, Mr. President, nothing at all," said Hart. "The Colonel was instructed by Henderson to find the communications equipment and set a contact schedule—hourly if possible."

"What if the equipment isn't available—or the Russians found it first?"

"That was covered. If he can't get to the radios, he has to use public or private phones, which will be scrambled. Meanwhile, Henderson's been in contact with his troops at the Guard camp about 160 miles southwest of Tower-Soudan. They have state-of-the-art communications gear with them, helicopter-loaded and lifting off the minute there's a break in the weather. The Commandant's also ordered a Harrier squadron from Cherry Point to Ripley. Quinn will have the jump jets, loaded with rockets, missiles and napalm at his disposal within the next ten hours—providing they can find the same break in the weather and . . . we can end run Stingers . . . ."

"How about the SPETSNAZ Tom? Do we have any input on them? How did they get into this country? Or get orders out to the Guard? That's scary as hell—the enemy operating inside your own country—inside your military!"

"General Smallzreid's been in touch with his people during the break. Burton should have something to report when we reconvene." Hart looked into the Situation Room.

"Ready for Round Three?"

The President shook his head in the affirmative but put his hand on Hart's forearm.

"One last question: How in hell can we keep a lid on this thing? Somebody's going to get wind of it. If I can't even communicate to my own wife . . . ."

The glum-looking NSA shook his head. "I'm trying Mr. President. It's calling for mirrors, piano-wire and some help from above."

*   *   *

Mordecai Baratz had the information he'd set out to obtain—and a promise of a rendezvous with the comely staffer in the Press Secretary's office. Cohen's project took longer but she was able to work her way into a three-way conversation with the First Lady upon her return to the

71

party. By 10:45, Baratz, by way of a pay phone on a damp Washington street corner, relayed the information to the Israeli Embassy. Speaking in rapid fire Hebrew that included arcane and archaic words over a line always scrambled, General Baratz informed the Mossad watch officer that the American military and political hierarchy was involved in non-stop, top secret sessions—meetings which had started around noon. Within minutes, a Mossad agent left the Embassy on International Drive and from another pay phone miles distant, made some late-evening calls that put paid and unpaid persons on special alert. Several received immediate assignments.

*   *   *

"They'll be armed and SPETSNAZ. More than two and this could be the end of the road!"

Quinn looked down at LaMont. The thick sweater was gone. A thin cotton shirt, opened at the neck, exposed his skeletal condition. Every device to call attention to his physical condition had been employed. Karin adjusted the dextrose bottle hanging suspended over the scientist's left arm, placing the oxygen mask in his lap.

"Karin—you're OK with this charade?"

In the dim light, Karin Maki brushed blonde hair back and recited her new assignment—"I'm the nurse's aide from St. Louis County Health Department—come in once a day to clean, change and give the Doctor his pills—hook up the IV's—straighten up the house and prepare the next two meals. Tonight, the storm is causing problems. If I leave, I may not get back and the doctor could be in trouble. He's in trouble anyway because we've only got 24 ounces of dextrose—ten hours supply and a half tank of oxygen left. I'll stay the night and ask the sheriff's department to deliver emergency supplies in the morning."

Quinn grinned. "You're a quick study—you'd make a great Candy Striper . . . ."

A smile broke over Karin's face as they shared the private joke.

Stepping out of the shadow, Quinn knelt in front of LaMont. He looked at the man's haggard face, up into Karin's eyes and back to the doctor. "You're ready to face the Russians?"

Quinn saw the white, bony knuckles flex slightly, the Adam's Apple slide up, a twitch of the head: His answer.

Rising, Quinn checked his watch.

"Ten minutes on the nose. I'll watch from over the garage. Be back here when they show up."

*   *   *

Standing in the dark on the glassed-in porch, Quinn felt for the second clip in his slash pocket then glanced again at the digital watch. Pale yellow numbers indicated slightly more than fifteen minutes had passed since the call from the mine.

Blowing snow limited his vision to the edge of the driveway. Tall pines formed a barricade between house and road. Peering almost straight down, he checked for footprints. Howling winds had obliterated Saatela's tracks. Quinn shivered in the uninsulated room, used for summer sleeping. White fingers of wind-blown snow streaked through gaps near the floor. A sudden glow of light caught his attention then disappeared behind the bending pines.

Straining to hear an engine, he saw a shadowy form at the base of the largest pine guarding the driveway. It advanced half the distance to the house, paused and turned. Quinn caught the motion of the man's arm signaling a backup to come forward. Edging into the corner he watched the second form plow through the snow, machine gun in hand. A beam of light suddenly appeared and Quinn backed away. An eerie, pale halo effect was created in the window as a light beam worked its way from the opposite end of the second story porch to his corner. It played the length of the room. When it disappeared, he edged back. The dark forms were almost directly beneath him. Quinn stared at the edge of the road for other shapes but saw none.

Time to go. I can't wait to see if they sent more than two men to pick up a cripple. Moving quietly, he returned to the living quarters where Karin and LaMont waited.

He raised two fingers and paused. Karin's hands were clasped on LaMont's shoulders. If things turned sour, the .22 in the canvas bag hung behind the wheel chair was her only salvation. Quinn knew a SPETSNAZ with a machine gun would have her splayed against the wall, her body a bloody sieve before she could act.

Quinn retreated into the small guest bedroom, Luger held close to his parka, vision limited to a narrow spectrum of the room—Karin and LaMont. By pre-arrangement, Karin would take her hands off LaMont's shoulders if someone moved in the direction of the bedroom. Seconds

73

seemed to pass like hours as he waited. *What's going on in Karin's mind? Or LaMont's? The doctor's going to lose control of his bladder when he looks down the barrel of a machine gun. Karin's been Miss Cool but . . . .*

Quinn tensed. He heard a slight sound and could tell Karin and the Doctor were seeing the face of the enemy for the first time. The door leading into the room from the stairwell had been left ajar. No need to make the prison chasers break it down.

"We did not expect company, Doctor LaMont."

The flat voice, virtually a monotone, sounded much like the Colonel on the highway. Quinn was sure, however, that he wouldn't have risked leaving the ORCHARD to retrieve LaMont.

"This is Miss Maki—nurse's aid. She comes every day. Tonight, she may have to stay . . . she . . ." LaMont coughed hoarsely.

"*She*, doctor, comes with us. My orders are to bring you to the mine."

Quinn kept his eyes on Karin's hands. They did not move.

"But Doctor LaMont can't be moved!" Karin's voice was strong, authoritative. "He has AIDS. Everything he needs to survive is here. He can't risk being out in this weather much less down in the mine. I don't know who you are but Doctor LaMont is *my* patient. He stays!"

From the silence, Quinn guessed that the soldiers were caught off guard. The reprieve, he felt, would be short.

"It is general knowledge the Doctor has a disease—multiple something or other." The disembodied voice had lost some of its earlier sureness.

"Multiple Sclerosis," LaMont said quietly, "was chosen as a cover because the symptoms, at this stage, can be very similar to AIDS. My diagnosis of Kaposi's sarcoma was made nine months ago." LaMont's right hand fumbled in the side-pocket of the wheel chair.

If they bite—they bite now! A smile filtered across Quinn's face in the darkened room. The leper was playing his role to perfection.

LaMont retrieved the small vial, offering it to his would-be captors.

"Here take a look. This is an experimental drug from France. It offers the best hope for people with human immunodeficiency virus."

Quinn saw LaMont's match stick-thin arm move up and down.

74

"Take it!" ordered Karin. "Look at it! It's the only drug to help people in this stage of the disease!"

LaMont's hand came back into view; bottle still clutched in hand. *The bait was taken. Come on Karin. Doctor. The piⱢce de rⱢsistance. Now!*

Doctor LaMont's head snapped forward, a gurgling rattle in his throat. A hawking sound followed and Quinn could see Karin slapping the Doctor on the back with her open palm. With her other hand, she placed a wad of tissue over his mouth. When the coughing ceased, she flipped her wrist in the direction of the soldiers, exposing cottage cheese tinged with curry powder—and drops of ketchup. She dropped the wadded tissue in wastebasket, her hands coming back to their original position on the wheelchair.

Quinn heard muted conversation.

"Our commander is not aware of the gravity of the Doctor's condition. The sergeant stays while I return to the mine. You are to remain in this room, answer no phones and do absolutely nothing until I return."

The clipped tones removed any doubts Quinn had about the identity of the speaker.

"A bag Miss!"

What now? He eyed the narrow opening. Karin had produced a small baggy from somewhere and was taking the vial from LaMont.

"This is all we have." Karin handed the bag to one of the men out of his line of sight.

"And this is the only proof I can bring my Commander. If the good Doctor is gravely ill and cannot be moved, you will spare it for that reason will you not?"

Karin nodded—hands again on the back of the wheelchair.

Quinn lowered his head, straining to hear the movement in the room. He heard one set of footsteps retreat down the staircase. *What in hell is the sergeant doing? . . . If he's convinced he's in the presence of an AIDS victim, he won't get too comfortable . . . .*

"Don't do that."

Quinn's eyes snapped up, the Luger rose higher in the air. Karin's hands gripped the chair.

"Why not?" came a gruff reply.

75

"Oxygen for God's sake! And if you're going to stay around here, getting in the way, point that machine gun down and shut the door. Heat's disappearing into the stairwell."

The momentary silence was shattered by the phone. Through the sliver of space, he saw Karin turn only to freeze at the command. "Do not answer!"

Quinn's palm grew sweaty. The ringing continued.

"That's my mother. She always calls to see if I've arrived safely."

Karin's hands came away from the ailing man's shoulders.

Quinn tensed. *Don't push Karin. Be cool.*

The phone continued to ring.

"She'll start to worry if I don't answer. She'll ask the police to stop by."

Darkness filled the narrow slit. He could see the mottled green of a winter battle jacket and realized the guard had positioned himself between the bedroom door and Karin. *He's getting nervous, doesn't know what in hell to do. But can I be sure this guy's SPETSNAZ?*

"If the police come, who do I tell them you are? Why would a National Guard with a machine gun be up here? Do you want to explain? They'll be here within minutes—the station's only a few blocks away."

"I, I am not Guard. I am regular army."

The choppy response was Quinn's clue. Pushing open the bedroom door he jammed the Luger into the man's exposed neck below the helmet.

"*Dhastrovya* Soldier!"

"Eh?" was the startled reply.

Right arm slamming down on the soldier's right wrist, he drove the point of the black grease gun downward. A burst of gunfire splintered the floor—the last round tearing a hole in the man's boot. The M3 clattered to the floor. Swinging left to confront his attacker, oblivious to the gun at his neck and shattered foot, the muscular soldier tried to face Quinn, helmet falling in the process. Quinn rolled with the man, letting the nose of the Luger slip from the neck to the shoulder blade and then pulled the trigger.

76

The arms of the SPETSNAZ shot straight out and the man plunged headlong onto the overstuffed chair, crashing against the wall face down.

Karin and LaMont stared at the fallen soldier. The room reeked of cordite as the phone continued to ring.

Picking it up, Quinn handed it in silence to the scientist who swallowed hard.

"This is Doctor LaMont." He listened for a moment, his gaze first to Karin, then to Quinn, tension fading from his ravaged face, replaced by a growing smile.

"Please. Yes. That will be just fine. Thank you."

Handing the phone back, his eyes were damp.

"Our paper boy. Do I want the driveway shoveled when the storm's over?"

*   *   *

"We've got to patch this soldier up. Alive, he might be of some use." Quinn motioned to Karin and together they pulled the man off the overturned chair, laying him out on the wooden floor. Karin felt for a pulse.

"Probably in shock, Caleb. He might be losing blood."

"Check the eyes, open his battle jacket and get the wound exposed. I'll get his boot off. Find out how badly his foot is damaged."

Working quickly, silently, they cut bloody cloth and leather away from the man's body, Quinn with a K-Bar and Karin, her *puka*, a small, lightly curved, razor-sharp Finnish knife that appeared from a small sheath taken from a chain around her neck.

"Can I do anything Colonel?" asked LaMont hoarsely.

Quinn looked up. "I'll get you in front of the communications board. Try to contact Washington and hope they can scramble at their end."

Pushing up from his kneeling position, Quinn aligned LaMont with the hidden radio and returned to the task of removing the soldier's bloody boot. Karin cut through the wool sweater covering the man's torso.

"You're sure he's not American Caleb?"

Gray-green eyes expressing a hint of angst, Quinn glanced up. "Not a hundred per cent, but his conversation was jerky—probably the pressure you had him under."

"Oooh Caleb. the exit wound—it's the size of my fist!"

Sliding next to Karin on her heels, Quinn eyed the ragged hole. "Big, but not deadly. Bone chips to remove and maybe some repair work on the muscles. Clean it up the best you can, stuff it with clean rags and tape it for now. We'll roll him over and patch it. About twenty-five years old, he should make it OK. Two toes gone but the bleeding's stopped." Quinn leaned back, wiping his hands on a towel, eyeing the man and the hole left by the bullet. Suddenly, he bent over, his left hand coming to rest at the edge of wound. Without looking up, he asked, "Your knife Karin!"

She handed him the *puka*, dropping on her knees to his side.

"Is there a problem?"

Quinn shook his head, eyes intent on the point of the blade as it touched the exposed red muscle beneath the top of the open wound. Pressing with the fingers of his right hand, he squeezed the object between muscle and epidermis. Gripping it carefully by the round, exposed edge, the wafer was easily extracted. Quinn held it in the palm of his hand. Dull silver, almost the color of pewter, the size of an old fashioned fifty-cent piece but seemed to weigh more than a regular coin.

"What is it Caleb?" She bent forward.

"Don't know. I never saw anything like it." He wiped the object, handing it to LaMont who held it up to the light and shook his head. A signal tone from the console caused him to turn his head.

"The signal—Langley's verified my entry code." LaMont looked over his shoulder. "Do you want me to go on?" He glanced at the small, circular device then placed it on the edge of desk.

Quinn wrapped the man's foot in a bath towel and rose.

"Punch your response number," he ordered, taking the headset from the doctor who leaned forward, pressing the miniature keyboard. Adjusting the mike, Quinn waited until the verification light blinked green, disk momentarily forgotten.

"This is GATEKEEPER, who've I got?"

"ROCKAWAY ONE Colonel—nice to hear your voice again."

Quinn, surprised, looked at LaMont who shook his head from side-to-side.

"ROCKAWAY ONE, have we met before?"

"We're on laser scramble Colonel, feel free to talk. This is Mumford. You asked I be assigned to you."

"Mel! Good to know you're in the loop."

"Not just me Colonel—you're feeding into the SR simultaneously. They need an update and I need marching orders."

Quinn smiled. He could picture the pudgy, brainy, Melvin Mumford, Russian analyst and farm boy from North Dakota.

"Top priority Mel. Build up a computer program showing every low-pass satellite coming within reception distance of this location. Include *all* birds in geo-synchronous orbit you could reach from here if you had the right transmitter—including cold ones. Sort'em by national origin, date of insertion, known mission. Is it operational or dead? I'm guessing the data's gonna be transmitted to Russia via satellite."

"S'cuse the interruption Colonel, but are you sure it's Russia who's gonna' get the output?" Mumford's question caught Quinn off guard.

"I've got what I think is a wounded SPETSNAZ, laying at my feet, Mel. Is there another possibility?"

The intelligence analyst laughed. "You know my habit Colonel. Looking in all the corners before pointing fingers . . . Crow's Maxim . . . remember?"

Quinn smiled. "I remember. But back to the satellites for a minute, once you've got that matrix made up I need to know what options exist to destroy a given satellite or—render it inoperable."

"Damn, Colonel, that's a tall one! There are hundreds of birds out there including junk. But remember the Kennsington Kids in England?"

"How do they figure in this?" Quinn scratched at his now stubbled chin.

"We've patched them into our system. Their ability to identify new satellites—now birds from other countries as well—is phenomenal. They'll help I'm sure. What's next on your shopping list?"

"Caleb—take a look at this." Karin had turned the wounded soldier over on this stomach, cutting through the material to expose the entry wound.

79

"Hold a minute Mel." Quinn turned, looking down at the now exposed back of the young soldier.

"I'll be . . ." he whispered. Dropping to his knees, he traced the raised welts starting near the man's shoulders, crossing one another in the small of his back. The dime-sized hole made by the bullet passed through one of the ribbons of scars.

"What's happening Colonel? You still there?" Mumford's voice filled the headset.

"Still here Mel. I just got a look at the backside of our wounded Russian—a mass of scar tissue, like he's been whipped—frequently."

There was silence at both ends of the conversation. Karin's fingers touched the old wounds gently. She looked at Quinn, wonder in her eyes. LaMont stared down at the mass of tortured flesh.

"Caleb, Henderson here. We've been listening to your conversation with Mumford. If you're through with him, we'll take over. There's some information on the SPETSNAZ connection."

Nudging Karin, he pointed toward the wound. Quickly, she prepared the temporary bandage.

"Jocko, good to hear from you, but Mel, before you go, I've just taken a small disk from the shoulder of SPETSNAZ trooper, the size of an old fashioned half dollar—about as thick. No external markings and looks hermetically sealed. Got any ideas?"

"Let me bounce it around with my boys."

Quinn placed the disk back on the console's ledge.

"Caleb, give us your assessment of the situation," said General Henderson. "How do things look?"

"To be blunt General . . . not good! We've bought some time but I expect our Colonel Hartlett will be heard from soon. We've got a comatose soldier here who can't be interrogated and the other unusual thing about him is the condition of his back—evidence of multiple whippings. Weather's horseshit and going downhill. Two feet of snow on the ground and winds are creating drifts fence high already. No way at the moment to make contact with the tank force, which I suspect, is in place around the shaft entrance. Hartlett's men have the university people under their thumb and haven't hesitated to get heavy-handed with some of them."

"Any killings?"

"None we're aware of. One of the team leaders, a man by the name of Bochert was displayed on the screen pretty badly worked over. But he appeared to be alive."

"Caleb, who's with you?"

"Doctor LaMont and Karin Maki—same party I told you about earlier."

"Shouldn't you get her the hell out of there?"

"We covered that earlier, General. She elected to stay on the payroll as a combat nurse and LaMont's lifeline. Sorry, but I can't get along without her and neither can LaMont. I've also recruited a deputy sheriff with orders to find more men and some tools to work with. I doubt if weather will allow troops from Camp Ripley to make it here anytime soon."

"Your decision, Caleb, I'll back you all the way. Now, can you neutralize the situation? Can Hartlett be stopped?"

"Only if I can get into the mine and disable the computers. If that's not possible, the only way left is to keep him from transmitting." He paused, "General, you had something on the SPETSNAZ...."

"It's not much, Caleb, but General Smallzreid believes his intelligence people have identified at least six infiltrators in the National Guard Bureau in the Pentagon. Two are in custody, four on the loose."

"Two in custody—learn anything?" Quinn kept his eyes on the man lying at his feet, who was returning to near-normal color.

"Lieutenant and a corporal. Corporal's a woman and both are proving tough nuts to crack. Speak fluent English but stiffly—hang on Caleb, something's coming in."

Karin looked up, "Learning anything?"

Quinn shook his head.

"Caleb—it's Henderson again. Smallzreid called his people working on the suspects. The woman was given a physical inspection by doctors from Walter Reed before interrogation. Same thing—whip marks on her back. They've just examined the lieutenant—identical scars. One other bit of bad news Caleb, those black boxes—we quite sure you're looking at Stingers—six of them with a minimum of six rounds each. We're gonna have a tough time bringing in close-in-air-support if they are up and operational...."

Quinn's hand went to his forehead. *Jesus on the Cross! Will this shit never end?*

"Six by six eh Jocko?"

"Minimum Caleb—very minimum, probably packed as they were sent to
Afghanistan."

"Those sonsabitches back to haunt us huh?"

"Fraid so Colonel. Let your people know, do your best to take them out."

LaMont's hand reached for Quinn's wrist. He whispered hoarsely. "On the big screen-our friend Hartlett."

Quinn reached for the volume control, lowering his voice. "General—I've got to sign off—pronto—we've got electronic company from the ORCHARD. I'll report."

*   *   *

"Three prisoners and all have signs of being tortured at one time or another. I don't understand it." The President looked down the length of the table. No one responded. He turned to Doctor Chen. "We've got to continue—to finish the assessment and formulate our next move. If Quinn can't stop the SPETSNAZ our only alternative is high-powered diplomacy—backed up by military power. I'm sorely tempted to initiate the Hot Line but before I do, tell us, Doctor, why would the Russians or the Commonwealth leadership be willing to risk war over this information? If we can come to grips with that, perhaps our next move will be easier to arrive at. They've got more internal problems to handle than you can shake a big stick at—why this—why now?"

Chen flipped through the growing stack. "Light can be shed on why Russia or any other country would want the data being developed in DANTE'S ORCHARD. If you will bear with me . . . ."

The President, nodded, turning in his chair. Swinging his left leg onto an overturned wastebasket, he leaned back, massaging the swollen knee.

Chen adjusted his glasses: "CRAY computers, any one of them by itself is a wondrous machine. Doctor LaMont's genius was to plug an equation into one machine, and as the primary calculations were made, automatically feed them into the second—he called it making Cognac with computers: distill fine wine once and then distill it again. When one considers the myriad factors involved in the astrophysical aspects of a pole shift—planetary alignment, unusual sun-moon relationship—a passing celestial body such as a dark star—impact by a meteorite, planet or comet, solar flare up or the disappearance of the earth's magnetic field—the computational permutations boggle the minds."

The gray-haired Secretary of State leaned forward. "Disappearance of the earth's magnetic field? Is that possible?" his voice incredulous.

Chen pointed in the direction of the military. *They* can assure you of that Mr. Secretary. The earth's magnetic field strength is approaching zero and may reverse as early as the year 2030. Over the past twenty-five hundred years it's weakened by fifty percent alone. When it disappears altogether, it will presumably reappear with reversed polarity. This is also why military aircraft and intercontinental ballistic missiles operate with inertial guidance systems. The magnetic compass is a thing of the past for precision flight."

The Secretary looked down the table. The military hierarchy nodded.

Chen continued, "Considering the more likely possibility, what Doctor's LaMont and Theodopolus have focused on—the need for high-speed, entrained computers becomes even greater. Of primary importance is the increase in the overall size of the polar ice caps—followed by changes in the surface mass loading—induced changes such as erosion and reduced water tables. The magnetic field disappearing has to be considered a geophysical phenomena as well as astrophysical. In addition, there are convection currents in the core and mantle of the earth that are tied to earthquakes and volcanic eruptions. These factors are linked directly to Plate Tectonics referenced earlier."

Pausing, Chen looked around the room. The Crisis Management Team hung on his every word. The little professor went on. "And finally, the human part of the complex equation—major socio-economic considerations—atmospheric pollution, *greenhouse effect*, mining, drilling, damming and nuclear war, testing or accidents . . . like Chernobyl . . . not to mention the most recent problem—the Shamal Winds and what they carry with them—along with Pinatubo's sickening contribution."

The President's leg swung off the wastebasket, both arms coming to rest on the table.

83

"How does all of this tie together?" Chen asked of no one in particular. "Consider our early understanding. Scientists coined the word 'crust' and believe the earth consists of a liquid interior core with a thin layer at the outer edge—boiling soup with a layer of fat at the top. Today, thanks to technological tools like sonic waves and deep drilling devices such as Glomar Challenger, most scientists concur that the crust, although solid, consists of seven large plates and several smaller ones floating on the deeper layer. These move like huge blocks of ice in a river breaking in a spring thaw. Made of continents and ocean floor alike, they move independently in various directions and at varying rates—motion believed to be caused by convection currents of the molten rock—magma moving massive plates by friction. When these plates collide, the denser rock of the sea bottom slides under the continents, the edges pile up and form mountains. In brief, the process of Plate Tectonics. And how do pole shifts tie to Plate Tectonics?"

Chen pushed the glasses up his nose, sipped from his glass and went on to answer his own question. "There are three principal forms of pole shift mechanisms. The planet capsizes while the spin axis migrates in the opposite direction *within* the planet, two, the planet's crust slides *around* the interior—or three, the planet simply *flips* end-over-end because of weight imbalance. Taking present events into consideration, the last two options seem the most likely to occur . . . of the two—which one is most probable . . . I cannot say."

The geophysicist dabbed at his forehead. No sound save that of the "*white noise*" could be heard in the room.

"What are the triggering elements? Centrifugal force, gravity or electromagnetic? Perhaps one or more forces converging—working together. As stated earlier, it takes billions of calculations based on virtually millions of constituent forces to determine a starting point. Software and hardware now exists making these calculations possible." He paused, looking around the room, "they are 2500 feet beneath the ground in Northern Minnesota . . . ."

Matteo Chen pushed his notes together. "I've taken several minutes to describe what one must study over seven to ten years in order to claim to the title of '*geophysicist*.' With this general background, you can better appreciate what DANTE'S ORCHARD is about to produce."

84

His last statement faded into the white hum as he reached for another document. "Doctor LaMont perfected the software that could manage billions of calculations quickly—when a critical factor changed. He could make the CRAYs spit out the predicted result. But his numbers alone are meaningless to all but the scientist or mathematician. This is where young Theodopolus made his outstanding contribution. He took LaMont's numbers and brought them to life by making them dimensional. Like cartographers who first gave us our navigational maps, Theodopolus, through his genius in the realm of Computer Aided Design technology, wrote the program allowing creation of dimensional maps to show *what* the earth's land mass will look like *after* the polar shift takes place. This he termed the LAND MASS RESTRUCTURE PROFILE – or—LMRP."

Chen let his notes drop as the civilian and military leadership of the United States pondered the new acronym pregnant with negative implications.

Picking up the document, he continued, "And together LaMont and Theodopolus developed a computer program that will predict with a higher degree of accuracy, *when* that shift will take place."

The little doctor looked around the room. Faces grim, the men gathered hung on his every word.

"According to LaMont's report of four days ago, trial efforts proved successful. The full blown run was set to take place sometime late this evening, early tomorrow morning at the latest."

President and professor looked at one another in intense silence. The Chief Executive's face was drawn and ashen. Chen's voice continued low, almost muted. "Whoever, Mr. President, controls DANTE'S ORCHARD over the next twelve hours will take possession of the computer-predicted *clock* and—*calendar* of the next polar shift."

Eyes closing, head tilted back, the President tugged gently at the loose skin of his neck. When he opened his eyes moments later, he saw the pursed lips and somber faces—all turned in his direction.

Chen paused, dabbing at his forehead before continuing, "And, they will possess a detailed, graphic portrayal of what the world will look like *after* it takes places—destruction of old land mass—creation of new masses."

The room was tomb-like as Chen glanced at the President over the tops of his glasses. His gaze was locked on the Commander-in-Chief. The words came slowly.

"In short, that person, party or—country—will have *exclusive* possession of what the *new* world will look like. Most importantly—that party can see *where* survival is most likely to be achieved, and thus gentlemen, dominate the post-Armageddon millennium . . . ."

*   *   *

The Commander-in-Chief got his first comprehensive overview of the newest battleground—185 acres of rocky outcroppings, pines and miscellaneous buildings located on the southern edge of the Boundary Waters Canoe Area. On the screen, in the darkened conference room, Yoder pointed to the grainy photo.

"Thanks to the photo analysis boys at Meade, we have this infra red shot taken by the NRO's Aurora. No way a Stinger will get to that bird! And," Yoder glanced at his watch, "we'll two Gnat's operational out of Holman Field in St. Paul. A pilotless drone, we should be able to get highly detailed video before a Stinger operator can track and lock in." The optical arrow from his flashlight focused on a dark, angular object. "The Head Frame supporting the cable mechanism raising and lowering the cages is the only entrance to the mine's twenty-seven levels of tunnels, drifts and stopes as they're called—and the Project itself—on the bottom at Level 27. The cage is stacked—one compartment lashed to another. If you don't come and go on this device, you don't get in or out of the mine. Once down 50 feet, the temperature remains a constant 53 degrees."

From the far end of the conference the President asked, "Where's this information coming from?"

Yoder lowered the pointer, turning to the President. "A Doctor Symington from the University's Physics Department and the Park Superintendent, who was located on vacation in the Bahamas."

The arrow focused on a large, square object. "This is a key building in terms of power." Yoder took a sip of coffee and cleared his throat. "Electrical motors operate huge drums raising and lowering the counter-balanced cages—like San Francisco streetcars."

The light beam traced the cables from Power building, through a smaller tower and then into the sheave wheels.

Looks like some kind of bizarre church, thought the President.

Air Force Chief of Staff, General Benedetto, voice betraying his Brooklyn roots, spoke out from the military end of the conference room. "Mr. President, we could neutralize that building—blizzard or no blizzard. Two of our Top Guns from the 318th Fighter Squadron at McChord plus two radar-support aircraft are on the deck. A nighttime refueling could be arranged. One F15E Strike Eagle with a laser glide bomb could take it out within minutes!" Benedetto looked around the room marshalling support for his proposal. His peers were noncommittal. No one in the room doubted that the Air Force could deliver on its' promise—the electronic wizardry preceding and during the 100-Hour War loomed large in everyone's mind.

Hand under the table, rubbing his knee, the President gave a vigorous shake of his head. "Until we understand the situation more clearly and have a first hand appraisal from Quinn, *nothing* is sanctioned in the way of force!"

He turned to the CIA director.

"Andy, what else does PHOTORECCE tell us?"

The arrow bounced quickly from spot-to-spot. "Tanks and personnel carriers—all accounted for except one APC. Our Russian commander has the Head Frame and main buildings surrounded in what Smallzreid and Henderson agree—is a classic defense perimeter—two rings deep, tanks staggered and fire power interlocked enfilade-style and focused on the most likely avenues of attack. The switchbacks leading up to Head Frame offer excellent tank emplacement positions with natural fire lanes to boot!"

"General Henderson?" The President leaned forward. "Is our man Quinn familiar with tanks, perimeter defense and the like?"

Henderson cleared his throat, his words like tumbling rocks. "His unit, Mr. President, had the direct responsibility for the perimeter defense of Khe Sahn. *Any* asset available to secure the airstrip was at his disposal. Khe Sahn held. The role's reversed here but I'd trust his ability."

The voice of Admiral Montgomery broke the momentary silence that followed.

"From where I sit, that circular defense doesn't look quite circular—it appears no armor occupies the northwestern quadrant—any particular reason?"

Yoder swung the light to the area questioned by the Chairman. "This is a preliminary input from Symington who visits the site four times a year—the area referred to Admiral, consists of a deep hole mined out from the surface years ago. Behind it, he recalls, is a rough, stone outcropping. We have to guess it represents a natural barrier."

A low buzz interrupted the session.

Yoder reached for one of gray phones.

"And the lieutenant?" The question was followed by more silence. "Call back if they save him!"

Slamming down the phone, Yoder turned to his left, his voice resigned. "The corporal ingested cyanide. They got to the lieutenant—apparently had trouble extracting his pill from a back molar. The vial broke in his mouth. A minute amount of the chemical entered his body—he's unconscious and critical—nothing was learned from either one. They've X-rayed the woman. She has a disk in the same location described by Quinn. It's a safe bet the lieutenant has one too. They'll be removed and taken by helicopter to Langley so the lab can get at them. Tissue studies will be conducted at Walter Reed to determine if the torture marks give up any information."

The Commandant of the Marines leaned toward Yoder, "That means Quinn's got the only captive and he's comatose . . . ."

        *   *   *

Quinn looked at the SPETSNAZ colonel on the screen and back to LaMont. He swallowed hard. The geophysicist's face was ashen, covered with a sheen of sweat, his breathing labored. He turned to bring Karin back to the screen. At that moment, she fell back with a sudden shriek, arms flying up, legs buckling.

"Damnit to hell . . . what the . . . .!"

Yanking the Luger free, Quinn saw Karin on top of the wounded soldier—his good arm across her neck in a death grip. Mouth open, she gasped for air.

"Doctor LaMont!" Quinn's head jerked at the voice on the screen.

LaMont's hand fumbled for the oxygen mast. Suddenly, the phone rang. Quinn saw the radio console blink amber indicating an in-coming call.

Karin's body covered his target and was going limp as he stood helpless. The bandaged foot offered the only visible target. As Quinn's boot came down in a stomping blow, the man's legs snapped up and Karin's body slipped to one side. With one swift motion, Quinn delivered the toe of his Sorrel boot into the man's exposed crotch, smashing testicles against pubic bone. The soldier pretzeled, bile gushing from his throat, a deep moan emanating from his gut. Karin lay on her back, chest heaving, hands at her throat.

Quinn put the barrel against the young man's head, sorely tempted to pull the trigger.

"LaMont, Karin. Get to LaMont, for God's sake!" She struggled to her knees as the phone continued to ring, each sound burst seemingly louder than the last. The amber light of the radio console blinked steadily.

"I hear no response, Doctor LaMont, surely you haven't chosen to go out in the storm." The mocking voice of the SPETSNAZ Colonel rang in Quinn's ears.

"Get the mask on the doctor!"

Quinn angrily yanked the prisoner's hands away from between his legs and with one swift movement, turned him over. Lace from the man's discarded boot was used to tie his thumbs together in a tight knot. Then Quinn pulled the man face up. Legs still drawn up, the young soldier's eyes were wide with pain but they followed Quinn like a feral animal.

"I wish to communicate, Doctor LaMont!" The SPETSNAZ commander's voice was growing harsher.

Quinn whispered over Karin's shoulder. "Talk! Keep the SOB occupied with bullshit! Do what you can!"

She pressed the button.

The phone seemed to grow louder with each unanswered ring.

Quinn slid around, found the small clip hooking the house phone into the wall and released it. Silence, a sudden balm to nerves on the grinding edge. Sliding into the chair, he put on the headset, punching in his code.

"GATEKEEPER! Damnit, be quick." His voice was low, eyes darting between prisoner, LaMont and the Russian commander on the screen.

"Quinn-Yoder here. It's urgent we talk!"

"You got your priorities—I got mine! All hell's broken loose up here. LaMont needs help, the guy who owns the ORCHARD is demanding to talk and the prisoner just about killed Karin and some asshole was trying to call on the house phone. You've got thirty seconds!"

Quinn suddenly remembered his voice was being heard throughout the Situation Room.

"I'll keep it short, Colonel Quinn. You've got the only live prisoner and he's got access to cyanide—probably in his back molars. It's possible he can get it out with his tongue, faster if he can use his fingers. He's our only lead to the SPETSNAZ, so keep him alive!"

Quinn glanced at the young soldier, legs still halfway to his chin.

"That's it? Keep the sonofabitch alive?"

If you only knew, Jocko, how close I came.

"Caleb, we need information!"

Quinn disengaged the headset jack and turned to the moaning prisoner.

Slipping the K-Bar from its sheaf, he stepped toward the man, pausing behind Karin, now in direct contact with the SPETSNAZ underground.

"I'm the nurse's aide," said Karin. "Doctor LaMont's on oxygen. When he can talk, I'll put him on. If he starts to choke he might rupture something."

"I appreciate your problem, Nurse. I will stay on the line. Do everything necessary to help the Doctor." The Russian commander's voice was less threatening, almost solicitous.

Karin glanced at the screen, then to Quinn. When they looked back, a young oriental woman, arms bound, was standing next to the colonel. A second soldier held a Beretta under her chin.

Quinn pointed at the fluorescent table lamp. "Hold it over his head Karin." Voice low, he pointed to the prisoner and dropped to his knees behind the man, motioning the lamp lower with the knife.

"Wh . . . what are you doing Caleb?" The knife glistened as Quinn gripped the man's head with his knees and pressed the blade to the young man's lips.

Like a patient in a dentist's chair, the prisoner's eyes darted between those of dentist and the tool in his hand.

"Oral surgery. Hold the light directly over his mouth."

Hands tied under his back caused the man's hip to ride up, putting pressure on his throbbing testicles. The moaning increased.

Quinn eased the blade between the man's lips, making contact with the teeth. He kept his voice low. "When the blade goes up, soldier, open your mouth."

The prisoner tried to move his head but Quinn increased the pressure of his knees pushing the blade tip downward a fraction of an inch. A spurt of blood appeared but the man's lips remained sealed.

"Caleb! You're cutting him!" The light swayed in Karin's hands.

"Hold that damned light steady!" Quinn growled.

The man's sweating face glowed in the soft light. Fresh blood oozed out from under the shoulder wound. The lips remained pressed together.

Suddenly, Quinn reached over the man's mid-section and with his free hand, raised it in a quick karate-like chop directly over the groin-area. He achieved the desired effect.

"Nyet!"

The moment the man's mouth opened, Quinn pushed the knife in and gave it a quick twist, blade up.

"Go ahead son, bite the K-Bar—it's an inch and half of steel." Sweat dripped from Quinn's forehead onto the soldier's face.

Gurgling sounds rose from the prisoner's mouth as saliva mixed with blood drained down his throat.

Probing with his fingers, Quinn felt the back teeth of the lower jawbone for an abnormal protuberance but found nothing. He peered into man's mouth and saw what he was looking for. He asked for Karin's puka.

Using the tip, he tapped at the off-white tooth. It moved slightly. Quinn pulled the smaller blade back, using his fingernail to finish the task. Keeping the man's head locked between his knees, he eased the K-

Bar out of his mouth. Rising, he held the small object up for Karin to see.

"Bite into this and you're dead in seconds . . . the smell of almonds—without the joy—your last remembrance . . . ."

*   *   *

Major Stepan Blagou, Senior Watch officer at the KGB's primary listening station in the Western Hemisphere, Lourdes, Cuba, listened to the last of the assessment reports. Through the use of encrypted messages, supercompressed and sent in a split-second burst from ground-to-satellite and back-to-ground, signal demodulated and analyzed, the INTELL officer was ready to pass his report onto Moscow. He reviewed the incoming sources—the Russian Embassy on Mount Alto, one of the highest points of ground in the District of Columbia and it's line-of-sight access to the Pentagon, stations in Glen Cove and Riverside, New York, the Russian Consulate in San Francisco and AGI's—Auxiliary Gathering Intelligence vessels that looked like fishing trawlers but were, in fact, floating antennas off both coasts of the United States. Despite the turmoil among the Commonwealth republics, the intelligence apparat continued to function to some degree. Blagou was comfortable with his final re-write and pressed the button.

Twelve minutes later, 0824 Moscow time, the report was personally reviewed by Major General Alex Promyslov of the KGB. Promyslov was in the Green Room of the Kremlin making an early morning presentation.

The entire contents of Blagou's message, beamed through a series of satellites before down-linking to Moscow, had been picked off by a tramp steamer off Dakar, on the West Coast of Africa—at sea for months—and never seeming to deliver cargo.

At 0207 Washington time, the intercepted high-grade cipher transmit had been re-transmitted from the sea-going vacuum cleaner, "*bundled*" and fired up to the "Aquacade" system for relay to the nine-story building just off the belt line near Fort Meade. Processed through the Loadstone/Carillon system in seconds, it was relayed again, by microwave, to Fort Belvoir, all under the sheath of TOP SECRET UMBRA. At 0314, Yoder's phone rang and the condensed version of

92

the original Russian message was verbally given to the CIA Spymaster. He listened to the report in silence then hung up.

"Gentlemen, the Russian President and his top aides, are in the Green Room just as we are gathered here. The message from Lourdes indicates none of *us* are where we're supposed to be—we've been sequestered here for an inordinate period of time and they suspect some form of military action is going to come out of our meeting."

Yoder pushed a cold cup of coffee away.

"Damn right! We should be moving into a Red Alert!" Air Force General Benedetto slammed his fist on the table. Sugar cubes flew in several directions.

The President rose to his feet. "No action other than what's been approved! We've made the decision and, by God, we'll stick with it until facts tell otherwise. Do I make myself clear?"

His gaze swept the room. No one contradicted him by voice or look. His voice dropped.

"Look, we're all bushed. The press is already sniffing. If we can finesse our way through the day, maybe we can dodge the bullets."

The President looked to his NSA for assistance—something that would bolster his own flagging spirits. He rubbed his eyes, wondering momentarily how Kennedy "finessed" his way through the long hours of the Cuban Missile Crisis.

Hart responded quietly. "Too many bullets, Mr. President. If we stay cooped up down here, we're going invite inquiries from every quarter. I say we make a decision as to who appears in public, when and where. Shag people around, juggle schedules but get them out in the open and try to shut down the rumor mill that's in the crank-up stage . . . ."

* * *

The winds blew jets of cold air into the cavernous hangar, staging area for the airborne operation quickly being pulled together at the 53,000-acre National Guard camp. The training facility was just north of Little Falls, Lindbergh's Hometown.

Inside the partially heated hangar, enlisted marines in winter battle dress talked among themselves, the atmosphere was subdued.

All of the troops, marines undergoing winter training and Norwegian ski troops assigned the other half of the hangar, wore white

93

boots, body coveralls and parkas. Weapons and duffel bags, white skis attached, were lined up in orderly rows. Cigarette smoke hung in the air. The low hum of propane heaters feeding hot air into the engine compartments of the C46 Chinook helicopters, dubbed "Flying Banana's", a clue lift-off was near. Junior Non-commissioned officers tended to bunch up, comparing notes from Operation Desert Storm, while senior NCO's could fall back on comparable experiences in Viet Nam. When tech and master sergeant were ordered by the warrant officer to personally load body bags in a cargo chopper, conversation dropped to an even lower level.

*   *   *

Air traffic controllers responsible for the Duluth sector could, during the winter months, often work from late evening to early morning without handling any aircraft. Air Guard F-4's were limited to one night syllabus flight a month—the rest was daytime activity.

Sheila Graywolf, Fond du Lac Ojibwa, was happy for the respite from boredom. Standing in the tower, she stared at the driving snow and wondered why two Air Force F15's would want vectors to Duluth when a blizzard was taking hold of the entire lake head.

Working with military controllers located at K.I. Sawyer Field in Michigan and the Farmington ATC, Graywolf did her part in bringing down one of the aircraft in virtually zero-zero conditions onto a runway needing almost constant plowing to be called clear. She listened to the order go out for sanders to follow the plows—cost would be negotiated with the airport commission.

Graywolf saw the jeeps and fire trucks heading out toward the end of runway 9. The F15 Strike Eagle, twenty miles out over the wind-swept lake was on a straight in approach to Runway 27, announced it's intention to make a full stop landing.

Shaking her head, Graywolf grudgingly gave approval for the attempt.

Minutes later, the pilot announced "*on-the-deck*." and quickly advised his wingman to ask for vectors to another base—one offering better conditions. She didn't see the F15 until it appeared beneath the tower, preceded by a jeep, followed by the fire truck. She was intrigued by the unusual device hung beneath the sleek craft. It seemed too large to be a fuel tank—besides the craft had re-fueled in Denver as she

94

recalled from the clearance. Cocking her head as the entourage faded into the blowing snow, she guessed the length of the device to be as long as her Nett Lake ricing canoe.

*   *   *

Quinn, Karin and LaMont stared up at the video screen. Hartlett held the Japanese-American software expert by the arm. The woman appeared terrified, near tears.

"Nurse Maki, there is a Sergeant with you . . . I wish to speak with him." Hartlett pulled the bound woman closer, snapping her head with a jerk.

Karin swung away from the set, eyes meeting Quinn's.

"Talk—say anything . . ." he whispered hoarsely.

She turned to the screen and took the mike.

"*Your* Sergeant must not have a stomach for AIDS. He stood near the door while the other man was here. When he left, he said something about checking outside the house."

"We? Miss Maki. Explain."

Hartlett seemed to squeeze the woman's arm. She appeared to grimace in pain.

"Doctor LaMont and myself. And if I have any problems, I call Chief Heitala at home. He'll send the night patrol here in minutes."

Hartlett advanced toward the camera, dragging the woman with him. His voice was harsh."You would be ill-advised to call anyone. Find my sergeant! Get him on the speaker with you. NOW!"

Quinn pointed to LaMont, tugging at the mask. The sheen of sweat had dried, the eyes seemed brighter. He reached for the mike. Quinn nodded.

"The Doctor's taken off the oxygen. He . . . wants to talk."

"Colonel Hartlett, this is Doctor LaMont." LaMont paused, cleared his throat of phlegm and continued. "The soldier you sent—Miss Maki is right—he seems afraid of . . . ah, my disease. Nurse Maki *isn't* afraid . . . except that I might die if left unattended."

Leaning toward LaMont, Quinn whispered in his ear.

LaMont's eyes flickered, hint of smile in his eyes.

"My associate, Doctor Theodopolus, should be here any minute. When he comes, we can find your sergeant and Miss Maki can make arrangements for obtaining oxygen and dextrose. I'm short of both.

95

Unable to take much in the way of solid foods—IV, is the only means of keeping my metabolism going. Sustained exertion is difficult."

Karin and Quinn looked at one-another, hope in their eyes Hartlett could be kept at bay.

Hartlett seemed to be ignoring them momentarily, eyes directed away from the camera, nodding to someone out of sight. He turned to face them.

Voice softer, Hartlett looked up. "The sergeant is with you. And by the way, don't count on Theodopolus being at your side, at least for some time to come." The Russian smiled, seemingly amused at his own humor.

LaMont wiped at his eyes but continued to grasp the mike.

Quinn heard the groan and turned to see the young soldier struggling to see the screen. Quickly he placed his foot between the man's legs and bent low; "Raise your voice—we repeat the operation."

Eyes wide, the head swung from side-to-side.

Hartlett continued; "You will let me know immediately when he returns to your side Doctor LaMont, meanwhile, I have other business to discuss. You recognize the lady?"

LaMont coughed. "I do. Maria Kubedera . . . ."

About to continue, Quinn leaned forward whispering again. The scientist nodded in agreement. He offered no more information about the woman.

"We know she is a software designer, Doctor, your lead programmer. And, good Doctor, we have some specialists here of our own."

The camera swung to a group of people bunched together. The man called Mohacker covered them with his Uzi.

Quinn saw LaMont lean forward, a look of surprise spreading over his face. The geophysicist's head began to shake from side to side but he said nothing. The camera swung away to the white-haired officer in the American field uniform.

Hartlett's voice changed again. He pulled the young woman forward.

"She tells me, Doctor LaMont, it is impossible to continue. A *glitch*, as she calls it, has suddenly developed."

"Let me speak to Maria," asked the Doctor.

Hartlett's arm came around the young woman in a fatherly fashion. "Maria, this is Doctor LaMont. Can you hear me?"

Maria nodded.

"Sorry things aren't going well for you and the others. I understand your predicament but please—no heroics—cooperate with them. Do what they ask. Avoid bloodshed. Remember, our project deals only in numbers and theoreticals. There are no flags, no religions to die for— just numbers. Do you understand?"

Kubedera nodded, eyes blinking back tears.

Hartlett motioned off screen. A hand pulled the woman from view.

"Your words are most reassuring, Doctor LaMont. As soon as Miss Kubedera advises the problem has been located, you will be asked to determine immediately whether it is a man-made virus. If so, it will be solved by you immediately. If not, the Oriental dies. Do we understand one another?"

LaMont's bony hands tightened around the mike. Hoarsely, he answered "Yes." The cathode ray tube went blank with a spike of light fading to black.

Quinn turned to the soldier. "The sergeant represents another problem . . . if he doesn't show up, Hartlett is going to send another guard—or two—maybe more." He glanced up at Karin as he knelt beside the young Russian who stared at him through filmed eyes.

"'I know he is there'." Did you catch that from Hartlett?"

LaMont reached for the blood-flecked disk retrieved from the open wound. It slipped from his grasp, falling to the floor and rolled within the soldier's view. Quinn saw the man's eyes snap wide as it continued toward him, finally dropping on its side.

Reaching over, Quinn picked up the device, a smile on his face. Holding it between thumb and forefinger, he showed it again to Karin and LaMont. "Thanks to our young soldier's knee jerk reaction, I've got an idea what this unmarked coin of the Russian realm is all about . . . ."

          *   *   *

"Five more hostages . . . top ranked scientists plucked by these men and transported God knows how to Northern Minnesota. Four Americans and a woman from IBM's Haifa research center. Power Visualization System's specialists . . . a 'insurance policy' according to LaMont that

they can force the CRAY's to merge with the RISC chips . . . reduced
instruction set computing . . . they've got every base covered . . ."

Quinn shook his head as LaMont finished his background on his fellow researchers now down in the mine, guns to their heads.

*   *   *

Pale green bottles of mineral water, unopened, plates of baked delicacies, uneaten, rested on the green baize surface. Copies of the report presented by Major General Promyslov lay on the table.

The room had been silent since Promyslov, second-in-command to the KGB Director had left the private meeting room located directly above the Commonwealth's main conference hall.

"The evidence, Mr. President, is irrefutable. The President and his top advisers have been meeting non-stop for the past eighteen hours! A significant increase in communications traffic to and from their forces in the Far East and Europe is apparent. We have intercepted orders indicating the newest link to their underwater fleet has been deployed—months ahead of schedule."

"And that General Manarov is? . . . ." The Russian President looked over the top of his reading glasses.

The KGB Director reached for the sheaf of papers.

"TACAMO . . . militarized versions of the Boeing 707 have been ordered out from bases in Hawaii and Maryland. Twice the loiter capability of their old aircraft, these planes head for mid-ocean, lower five miles of trailing antennas into the sea and proceed to fly in steep banked circles which stabilize the wires in a near-vertical position. In this ingenious manner, they can issue global orders to *all* their submarines—something, regrettably, we have not yet mastered. Total and instantaneous communications access to their entire undersea fleet can only mean one thing—preparation for attack! In the Pacific, we've temporarily lost track of their missile-tracking cruiser codenamed, COBRA JUDY. We will find it—soon—and when we do I say it will be over the horizon from our ships based in Vladivostok!"

"And your thoughts Sergei?" The President swiveled in the black leather chair to face Sergei Korovkin, Chief of the General Staff and holding the rank of Field Marshall. By tacit agreement among the various republics, the Russian President would speak for them on matters of joint security.

A graduate of the Frunze Military Academy, veteran of the Hungarian Revolution, an architect of the Afghanistan Campaign, the outspoken hard-liner regarding the West, nodded, bulldog appearance matching his demeanor. As he spoke, his right fist was balled. "I say we activate the radar system at Krasnoyarsk—order the Carrier's Kiev and Koll to maintain stations in the North Atlantic by reducing speeds until supply ships can reach them. The Carrier Novorssiysk keeps its position with the pacific battle group even though it is having trouble maintaining full power. It Yak VTOL's on board would be needed to provide defensive coverage for the fleet should it come under attack. We have two new submarines fully armed with 16 ballistic missiles each on board at Petropavlovsk—order them to sea immediately. Rotation of submarines in the mid-Atlantic is delayed and we move our picket lines there another two hundred nautical miles further west. Additional submarines are ordered out from Murmansk and Kamchatka and the Caspian Flotilla is dispatched south to keep that area neutral." He paused, "This could be the opportunity that finally gives us access to Persian Gulf oil—cleans the slate of the Iraq debacle . . . ."

The Commonwealth leader raised his hand. "The moment we turn on that radar—without prior notice—*we* are in violation of the 1972 Treaty. Granted, it fills a vital gap in our defense of the Northeast quadrant, our most vulnerable, but we, not America will have sent the first *official* warning signal. My assessment is—virtually all of the American activities reported thus far are of an internal nature—with the exception of the TACAMO deployment. Would you not agree, Viktor?"

The KGB Chief, torn between loyalty to the Commonwealth leader and respect for the nascent power of the Marshall, nodded in sullen agreement.

Rubbing his forehead, the President stretched his arms out over the green table, palms down. "What is happening Sergei, Victor, *appears* threatening but there is the nagging doubt in my mind the Americans truly mean to take the first step toward annihilation. In light of their President's willingness to proceed with the stand down in missile aiming priorities and find the funding to help solve the problem of the deteriorating warheads in the Ukraine and elsewhere, I cannot see offensive action in the offing."

Glancing first at the bank of clocks, he faced the two military officers; "Soon it will be Friday morning in Washington. Ambassador Titov will request an urgent meeting with the President. Meanwhile, we use the "Hot Line", advising we are turning on the radar at Krasnoyarsk—but for test purposes only. This we do on our normal even-hour transmission. By giving advance warning, we negate the threat of a unilateral action they could label openly hostile. If this situation gets worse, we simply leave the monster running. As to the military moves suggested by you, Sergei, delay of rotation seems plausible but ordering more submarines out borders on the provocative, as does extending our undersea lines toward America's East Coast. That I will not authorize or condone! I am satisfied our posture is correct. We can withstand the pressure the Americans are bringing to bear."

Rising, the President nodded to the Marshall and KGB chief.

"I suggest we attend to our normal schedule and reconvene . . ." he glanced at the gold watch, a gift from his wife after his election, " . . . in three hours."

The President gathered in the sheaf of reports lying on the baize surface. All were marked, "OVERSHENNO SEKRETNO," in red letters—"TOP SECRET."

*   *   *

Quinn had Mumford on the line. "A shot in the dark, Mel, but see what the hardware boys can do with that theory."

"Sounds feasible, Colonel. The disk's been removed from the dead corporal. A doctor at Reed is objecting to taking the device from the lieutenant because he's still alive—says he'll file a complaint with Amnesty International."

Quinn stared down at the man who would have killed them all with one burst of the M3 grease gun. "Somebody better get that medico under house arrest or in the rubber room. Our SPETSNAZ commander is threatening to kill a young software expert if LaMont won't cooperate. Tell Amnesty to haul ass out here—save *our* people's lives."

Mumford chuckled. "We'll have the inside story on the disk within an hour, I promise. If it's a position-locating instrument—we'll have the operational parameters worked out. Anything else we find of interest will be relayed back pronto. OK?"

"Haste will be appreciated, Mel. Now, what about receptor satellites in space. Any inputs?"

Quinn shook his head as Karin offered him a cup of coffee.

"Bad news first, Colonel. 946 objects fall into the overall count." Mumford paused, "now the good news. NORAD in Colorado says 890 are positively known to be pieces of junk from boosters, jettisoned from space shuttle work or dying satellites—Russian *and* ours. That leaves 56 suspect targets. An analysis is being run on each item up there. Given another hour or two, we should have the list narrowed down by half."

"Don't mean to make your work more difficult, Mel, but LaMont's informed me the run in parallel was scheduled to start about midnight our time. Problems, legitimate ones, seem to have popped up in the software. Now I find out there were people in that APC being transported up here—top flight specialists from IBM's think tanks. Unless we get damned lucky LaMont's going to have to solve that problem and then it's green light for Hartlett. Do the best you can. Now, how about interdiction possibilities?"

"Your boss or the Company will handle that. I'm back to the big board. Look for my call on the disks ASAP. Signing off!"

Hoping for a brief respite from the barrage of data he was dealing with, Quinn accepted Karin's second offer of coffee. Sensing his tenseness, she stood behind him massaging the muscles around his neck.

"Caleb-Henderson here. Congratulations on the discovery of the disk. It might prove valuable."

Quinn answered the Commandant. "If it turns out to be a miniature transponder, General, it could be the key to getting inside the tank ring and down into the mine."

"Glad you brought that up Caleb. There's been some, ah, spirited discussion—about how far you're allowed to go on your own."

Quinn waited for Henderson to continue then suddenly realized Jocko Henderson, his friend was speaking, not the Commandant of the Corps.

"Not sure I understand, Jocko. Mission assignment four hours ago was clear and straightforward—at least to me. If there's a change in plans, I need to know before the bugles blow."

"Diplomatic considerations, Caleb, I won't bother you with the niceties. Major problem—the Joint Chief's feel we have all our eggs in one basket—and you're it!"

Frustration on his face, Quinn bit his lip and looked up to the ceiling.

"Caleb? You still there? Henderson's raspy voice rang in his ears.

"There's a strong argument, Caleb, for having you hold back until we have our ducks in a row—then we take on the SPETSNAZ."

Drawing a deep breath, he could envision Henderson's Hot Seat position in the Situation Room. "General, you know I'm one strong advocate of ducks looking good on the drill field—but to drag up another old saw—all the King's men ain't gonna put this Project back together if Hartlett gets the CRAY'S cranked and his delivery system locked on. Unless I'm missing something, that's the whole ball game, isn't it?"

"You understand it very well Colonel Quinn."

"Sir! I . . . ." Quinn knew the President might still be present but hadn't expected him to intervene.

"It's been a long day here in Washington and I'll be the first to admit Colonel, one's thought processes get fuzzy. You've got a hell of a job to do and I've been listening to the exchange between you and your Commandant. I want to put your mind to rest. *You*, Colonel, are my designated field commander. We'll handle the diplomatic side of things! Your first priority is to stop Hartlett from running the program. If that can't be accomplished, then keep him from giving it to his countrymen. Keep the loss of life at minimum but do what you have to do with what you've got at hand."

"Yes, Sir!" Quinn found himself standing at attention.

"I take it that clarifies the situation General Henderson—Colonel Quinn?"

The two men, twelve hundred miles apart, answered simultaneously.

"Yes Sir!"

*   *   *

"Whatdidya'say your name was?" The voice over the phone was guarded.

"Quinn, Colonel Caleb Quinn, United States Marine Corps."

102

"Jahah Mister Marine, you picked a hellavu' time to call an old miner. It's past midnight. The village idiot knows we're three feet into a blizzard. How'd you get my name anyway?"

Prefabricated reply ready, Quinn answered; "Touched base with Chief Heitala. He told me you're the most knowledgeable man in town when it came to the mine."

"Jahah! Mister Colonel. What can I do you for you?"

"Mr. Huhta . . . Nick . . . There's a problem in the mine. A number of people are down there. They need help. Access is—well let's say it's difficult."

"Them eggheads from the "U" got things screwed up with the hoist again? Man, why didn't you say so in the first place? What happened?" The tone of voice changed.

"Complicated situation, Nick. Can you meet me at Doctor LaMont's home in, say, fifteen minutes? Bring what you can that will give an idea about the mine—layout, electrical, shafts."

"Hell, that operation goes back to 1870. You're talking a lifetime of mining."

"That's why the Chief tagged you Nick. And by the way, if you've got your deer rifle ready to go—and any hand guns, they might come in handy."

"Colonel . . . ?"

"Colonel Quinn, Nick . . . ."

"Guns? There some kinda' problem down there?"

"Nick, I can't give details, but you're damn close."

"Jahah! Mother should be alive to hear this, Nick Huhta going into the mines with a gun instead of a pick. Fifteen minutes, Mister Marine!"

Quinn put the phone in the cradle, taking another cup of coffee offered by Karin.

"Saatela's grandfather going to come?"

Quinn nodded. "Sounds like a crusty curmudgeon, but he's coming. Now I'm hoping his nephew, his recruits, more SPETSNAZ and Nick don't all meet downstairs at the same time." He glanced at his watch.

Karin nudged Quinn, pointing in the direction of the geophysicist. LaMont, head back, mouth partly open, was snoring softly, one hand cradling the oxygen mask. "We talked while I gave him a sponge bath.

103

He figures there's no more than three months to go—maybe less. All of the symptoms have increased—extreme fatigue, sweating, coughing fits, blackouts, fevers, lesions and worst of all—he said—depression."

Quinn's eyes narrowed. "Did he talk about the blackouts?"

"No," Karin replied, shaking her head. "Only mentioned he could tell they were coming by the onset of heavy sweating. He thought it was a chemical reaction to the loss of body salt. Something else Caleb—his medicine?—it might be a placebo . . ."

Quinn's head jerked up—disbelief mirrored on his face.

"You're kidding! One of this country's most important projects—he's a key to the whole thing and he might be on sugar pills? . . ."

Karin's eyes were misty. "He said he was a participant in a project that included U-of-M Medical Research and someplace in Atlanta—I forgot the name—its called Protocol 019. Four hundred AID's victims are involved—half are getting a second generation drug called AZT—the others . . ." Her voice trailed off.

Quinn's gaze shifted from LaMont, sleeping fitfully, to Karin. "How far along is the study?"

"Half way," she answered quietly.

Putting his cup down, Quinn looked at Karin. "Just what we need—another problem. If he's on the wrong side of the test we've got find out and get medication to keep his body and brain functioning. When he wakes up, ask him who's in charge of the study."

Quinn's eyes swept the room.

"Wonder if he keeps any weapons around—our armory is stretched pretty thin . . ."

Karin stroked LaMont's hand. "Wouldn't waste much time Caleb—he's a Quaker—his friend—a pacifist."

Head shaking in frustration, Quinn's lips pressed tightly together.

Karin's index finger came up to her lips.

"You hear something?" Quinn rose quickly, hand to the Luger. "I'm going to check from the garage porch."

Seconds later, they stood in the dark, looking down into the blowing snow. A fresh trail led to the garage. He knew it was more than one person. He pointed to Karin's twenty-gauge, then to the lights. She backed against the wall, turning off the switch.Positioning the flashlight under the barrel, Quinn lowered the weapon and with the tip, nudged the door leading down to garage open.

A dark shape blotted out part of the window.

Falling to his knees, then onto his belly, Quinn eased the gun over the lip of the stair, inching back to allow only his head to show over the top of the stair. A draft of cold air funneled up from below. In a second, the draft became a blast as winds found the opening. Quinn snapped on the flashlight pinning the young deputy, in semi-crouch, Magnum drawn.

"Saatela! It's Quinn up here!"

Saatela straightened up. "I've got forty pounds of high-grade dynamite, a dozen bed sheets and two good men. Dynamite's in your vehicle—men are with me."

"Bring'em up!" LaMont's eyes opened in surprise when Saatela reappeared followed by a bear of a man in a plaid Mackinaw and worn, turn-of-the-century lumberjack hat. He was followed by a dark-skinned, narrow-faced man in a black snowmobile suit and boots.

The man in the lumberjack hat spoke first.

"Lubo . . . that's love'—in Croatian—Lubokowski."

Snorting, the burly giant eye's glistened. "Should have known it was you when Wally showed up in the toy truck of yours. Couldn't remember where the hell I'd seen it before. Full bird colonel, eh?" The big hand came out of a chopper mitt, engulfing Quinn's.

Saatela motioned to the dark skinned, thin-faced man in a solid black snowmobile suit, standing silently by the door. Quinn saw the vintage lever-action carbine palmed in his hand, tucked almost out of sight along the leg of his suit.

"Colonel, meet Jimmie Running Fox. That's his name on the reservation. Off, it's just Jimmie Fox." The Indian's head bobbed slightly but he said nothing in the way of a greeting. Looking the man straight into his deep-set eyes, remembering the casual, but familiar way he carried his weapon, Quinn knew better than to measure this man by the strength of his grip.

"Jimmie's my *niita*—brother-in-law," said Saatela, zipping open his blue winter police jacket. He nodded toward Running Fox and spoke something unintelligible to the man who silently turned and

105

disappeared down the staircase to reappear moments later with two large grocery sacks full of folded sheets.

"Didn't sound Finn, Wally—you and your brother-in-law speak some other language?" Quinn looked at Saatela quizzically.

Lubokowski laughed. "Them two buggers speak teepee talk all the time—drives everyone up the wall. Can't tell when they're calling you a shitface or insulting the size of your weenie."

Pushing past Quinn, Lubokowski glanced over his shoulder. "Whatawegothere?" This sucker looks like he's in a world of hurt."

Quinn nodded. "Background shortly . . . introductions now . . . Lubo, Jimmie, meet Karin Maki and Doctor LaMont." All four acknowledged each other's presence with a casual wave of the hand. "We're operating on a short fuse. Lubo, give me handle on your background—I know you run some kind of lumber operation."

Peeling off the other leather chopper, Lubo took the coffee offered by Karin. "Been pounding pulp for five years. Before that was with the Seabees. Construction battalion. Ten years."

"Ten in Lubo. Why not twenty?" asked Quinn, already guessing the answer.

"Up and down the ranks like a yo-yo. After ten, my fitness report was about equally balanced plus and minus—figured it was time to departevoo while I was even." Eyes laughing, Lubo downed the scalding coffee in one slurping swallow.

Quinn turned to Running Fox.

"And you? . . . ."

Running Fox leaned against the doorsill, carbine tucked out of sight.

"Graduated from reservation school in '61, attended St. John's University. Degree in English lit. In to the Navy as a SEAL. Tour in Vietnam—out in '70. Back on the reservation teaching English."

Quinn caught the look of pride in Saatela's eyes.

He turned to the bound man. "This young Russian trooper was, up to a couple of hours ago, one twentieth or so of our problem." He looked at Saatela. "And speaking of head counts, I got your granddad on the phone—gave me a ration of shit but said he'd come. You ready for that?"

106

The glance Saatela gave Running Fox and the frown crossing the Indian's face was caught by Quinn. Looking down at the floor, Wally answered softly, "'Bout as ready as I'll ever be."

Karin tapped Quinn on the forearm, "Somebody wants to talk." The in-coming transmission light blinked amber from the corner.

"Stand easy. Nick should be here shortly and I'll cover things once for everybody's benefit. Sliding onto the kitchen chair, slipping the headset on, Quinn quickly entered his code.

"GATEKEEPER on line. Go!"

"Colonel, Mumford here. Two items of interest. The disk is what you suspected—and more. Not only is it a transponder; It's a form of under-the-skin dog tag. The woman soldier was Natalie Schevernchenko, Sergeant First Rank, *Voiska Spetsialnovo Naznacheniya.* Radio communications operator."

"*Voiska* what? Say again, Mel," asked Quinn, confused.

"SPETSNAZ . . . the corporal who committed suicide was definitely SPETSNAZ."

Quinn glanced quickly at the man lying on the floor.

"What else?"

"Back to the second disk. It's out but because the man died as well. We've determined the lieutenant was a disinformation officer and political assassin."

"That's it? Nothing else?"

"Micro chip records show both parties saw duty in Afghanistan and both appear to come from somewhere in the Ukraine. Doesn't mean a whole hell of lot—thousands of SPETSNAZ were posted to Afghanistan because that's where the action was and the Russians wanted their elite bloodied."

"How does the disk work as a position-locating device?" A hazy idea was forming in Quinn's mind—too hazy, he thought, to have any merit yet.

"Just like it does on an airplane—emits a low signal which limits its range. We don't have the exact distance parameters but the boys say it paints a signal on S and J-Band radar within 3,000 yards. Under combat conditions that could be adequate for a company commander to keep tabs on his troops. But as for the satellites, we're down to six suspect targets. Three are satellites. Two Russian, one commercial—a weather bird put up recently by the French for a private concern in South America. I'm about to push that off the list in favor of the two Russian satellites—same north-south orbital configuration and carried on the books as similar to our TIROS-1 weather satellites. We've got some electronic interrogation stuff going and I should have them under the spotlight shortly. You ready for Henderson?"

"Ready Mel."

Weather birds. Low pass. North south orbits. Makes sense. They could be up for months. Just putzing along, then press a button and bingo.

"Caleb, General Henderson here. You keeping the tent ropes tight?"

Quinn laughed at the familiar question.

"Yeah, Jocko, tight as can be. I've got my local laundry list completed and—a few good men."

"What are your needs, Caleb—that we can fill?"

"Several General, a good weather update for this area. I don't have time to listen to the local radio. Two, a complete run-down on the number of men involved on the National Guard roster up here and the exact number of SPETSNAZ we have to deal with. It's too early to start keeping a kill or containment tally but if I know the high side number, it'll help to check'em off one-by-one."

"That's a done one. Next?"

"What's happening in the satellite sector? How about taking them out of operation?"

"We have an F15 on the deck at Duluth with an anti-satellite missile loaded with smart rocks—the second was vectored over the storm cell to the Air Force base at Minot, North Dakota."

Karin poked his arm. Quinn glanced up as she whispered in his ear.

Glancing at LaMont, he was silent for a moment, then nodded. "One last item General. It's possible LaMont's on placebos instead of real drugs to handle his AIDS. We could use some help in getting him some of the real McCoy. He's in bad shape and suffers from blackouts he says are increasing in frequency and duration."

Lubo and Running Fox were staring at LaMont. No one spoke.

*   *   *

Wrapping the thick towel around his waist, the President eased onto the bench of the combination bath and steam cabinet. Whorls of steam rose from the opposite end of the small, glass enclosed room. The pleasant smell of eucalyptus oil tinged air and senses. Head back, eyes closed, his shoulders sagged as the room filled with the welcome moist heat. Quinn's image came to mind. The Commandant made his personnel jacket available for review by the CMT. The President smiled. *Smart move on Henderson's part. Joint Chiefs were hard pressed to find fault with Quinn as the battlefield commander.*

Legs stretched out, he reflected on the situation. *Quinn's in the barrel because he was forty miles from Tower-Soudan and the Commandant knew it. Virtually nothing to work with and the clock's ticking. How in hell can he pull it off? . . . .*

The glass paneled door to his right slid open. His wife, in her silk kimono held his favorite bathrobe in hand—a thick, well-worn terry cloth that reached to his ankles. The choice of kimono, a signal a truce was being declared. Startled, he looked up into his wife's eyes. "Tom Hart's on the phone. Said it's important—and confidential. He was kind enough to tell me you've got a major problem. Take the call in here— I'll close the door." Clutching the kimono near the neck, the other hand resting on her hip, she bit her lip. "If it means anything, I'm sorry for the way I've acted tonight."

Rising through the mist, the President ran the towel over his lanky body, dropped it to the floor and gathered his wife in an embrace, arms tight around her shoulders. He held her in silence for several long moments then tilted her head up. "I'll take the call—then we talk."

She squeezed his arm and left.

He settled in the chair, towel retrieved, and lifted the phone.

"Tom?"

"Mr. President, the stew is starting to thicken. Some is sticking to the bottom of the pot. High-grade intelligence data being picked up by all channels has tripled. Moscow, it seems is gathering decision-makers into the Kremlin from the various republics. The Northern and Pacific fleets have been given new orders that we haven't decoded yet. Caspian Flotilla has been told to standby for orders that would have them headed south toward Iran. No ships are heading back to port and those *in* port are being pushed to make ready for sea. New orders being issued to field commanders throughout Europe. Our LaCrosse orbiter, responsible for watching the activity at Baikonur Space Center has picked up photo and communications inputs indicating things are accelerated. At Tyuratam, an Energia rocket is being hustled for launch. We don't know the payload."

"How is all this being interpreted by the Situation Room?" he asked.

"With distinct nervousness, Mr. President." There was a pause, "and we've abandoned the people dispersal plan. Things are happening too fast. All the troops are back in the saddle down here."

"Give me fifteen minutes Tom."

"One last report, Mr. President. The Israelis seem to have their nose in the tent."

"How in hell? . . . ."

"Last night's soiree, Mr. President. The team of Baratz and Cohen was there. Yoder's people picked up information that seems to tie their presence at the party to priority messages to the Mossad."

The President's hand clenched the phone. "Who's responsible?"

"We're not sure Mr. President but we do know Baratz is going to work his charm on a press secretary staffer tonight and Cohen is going one-on-one with Air Force Captain Black—dinner for two at their sex-and-satin suite in the Watergate."

The President asked, "How did all this come about?"

"The staffer hustled, mentioned to a friend who mentioned it to the press secretary who reported it immediately to Yoder's safety watch. Captain Black reported his contact directly to Security. Both assignations will go on as scheduled.

Mr. President, there's one other small problem . . . ."

There was a momentary pause at Hart's end.

"The First Lady was involved." Stunned, the President stared at the fogged mirror."

"How in the . . . .?"

"When Cohen was spotted talking to Black, a floor agent kept her under close surveillance—when she worked her way into a conversation with the First Lady returning from your call, the agent put a directional mike into play. Cohen's taken more than Psych 101—she

110

knew the First Lady was distraught and in minutes had a pretty good fix on your recent schedule—appointments broken right and left."

"This sounds like Machiavelli in a Byzantine brothel!"

"Agreed Mr. President but considering what's happening up in Northern Minnesota . . . ."

Rubbing film off the mirror, the President sighed, studying his lined face. "Who's to say Tom, someone isn't watching me scratch my private parts this very moment?"

Placing the phone in the cradle, he flipped off the light and padded silently toward the master bedroom. His wife put down the slim book of poems by Wolfe and watched as her husband approached the bed then stopped.

"Temperature's rising down in the Situation Room. My mind is cluttered with crap but I've got to go back."

"Can you tell me anything?" she asked softly.

Shaking his head, dropping the robe at his feet, he pulled on his boxer shorts. "George Weller likened it to the Berlin Blockade, Cuban Missile Crisis and hostages in Iran wrapped up in one bad scene. Personally, I'd say this is worse!" Brushing his hair into place with his hands, he slipped on the tweed jacket. Bending down, he kissed his wife again. "Hold that last warm thought."

*   *   *

Quinn had the men gather in the living room. The meeting between grandson and grandfather had been strained—made more so when young Saatela introduced Running Fox as his brother-in-law.

Quinn's first impression when Nick Huhta clumped up the stairs— he was asking too much of an old man. Nick appeared to be as big around as he was tall. The iron grip partially changed his mind and he changed it completely when the old man pulled a weapon out of a sack proudly holding it over his head.

"*Suomi*—means Finnish—best damned submachine gun ever made! Kept it even though we were supposed to turn them over to the Russians. Screw'em says I! Means too much to me."

"You fought in World War II Nick?" asked Quinn.

"Hell yes! Twelve years old and making Molotov Cocktails out of *alkohoolikiike* mixed with kerosene, tar, and wrapped in a rag for a wick. Even made mines! *Talvisota* we calls it."

"Which means . . . ?" queried Quinn.

"A Winter's War," snorted the old Finn, "that's what *Talvisota* means . . . ." The words were spoken with obvious pride.

Young Saatela and his *niita* looked at one another but said nothing.

A gust of wind rattled an outside storm window.

Quinn gathered his thoughts. *It isn't fair to ask them to do this. They've paid their dues already.*

Nick looked at Quinn, a glint in his eyes. "Well Mister Marine you calls this meeting. People need help!"

"I'll get the heavy stuff out of the way, Nick." Quinn gave his new recruits an overall view, then paused to let the information sink in. "That's just the tip of the iceberg. Given time, I could generate a list of lesser problems."

Wally Saatela, blinked, looked around the room, then back to Quinn. "Have we got *anything* going for us?"

Quinn was silent.

"Jahah! For damned sure we got something going!"

All eyes turned to Nick, a wide grin on this face. "By golly, we got the weather on our side! For damned sure of that!"

*   *   *

Hunched over the table, Nick finished the crude sketch. Tourist folders from LaMont's file gave Quinn and his men another perspective as did photographs taken from a scrapbook maintained by Theodopolus. Using a cut-a-way view from a colored brochure, the old Finn described how the cage hoist worked and the miles of switch-backs and stopes—inclined in such a way that mules pulling carts could handle the load. The stopes snaked downward to the twenty-seventh level— location of DANTE'S ORCHARD.

Lubo whistled through his teeth. "Lordamighty!" He looked up at Quinn. "One way in—straight down—smack through a circle of—what did you call'em?"

"SPETSNAZ Lubo, Their very best."

Quinn tapped the sketch. "Nick, is that a fact . . . only one way in?"

Huhta peered up, clear blue eyes belying his age, laughed, head shaking.

"Here," he X-ed a spot northeast of the Head Frame. Alaska Shaft . . . built in '06 to improve air circulation as the levels worked lower. It collapsed in places as timbers rotted. They was never replaced."

"Is it passable Nick?" Quinn's finger came to rest on the X.

"Jahah, if you're a little batty!"

The men laughed as Nick continued.

"A few years ago, a boomer—somebody who comes up here in good times, when there's construction work to be done—hears about the Alaska in a bar—air flowing top to bottom. Makes a bet he could work his way down."

"And . . . .?" asked Quinn.

"Made it. Barely. Starts at dawn on a Saturday morning, half in the bag—shows up twenty-four hours later at the bottom, bloody, stone-cold sober and babbling to himself. Collected one hundred dollars and was gone. Last anyone saw of him."

"We don't have *four hours* much less twenty-four."

Nick poked the folder with the pencil. "Don't have to ride down *in* the cage."

"Meaning what Nick?"

"Twice a year, counter-balanced mechanism is geared down and one man, a volunteer, rides on *tops*—does an examination of the shaft's condition—loose rocks, timbers that needs replacing, condition of the safety ladder and the cable break safety rig. Hell of a deal! Hangs on for dear life with one hand, eyeballing the shaft with a high powered flashlight in the other!"

"You ever do it Grandpa? Wally looked at his grandfather.

Nick grunted, muttering, "Jahah! You ain't paid your dues as deep rock miner if you didn't do it at least once." He snorted. "After the union got stronger, management pansies had to do it themselves. More than one trip down ended with britches full of shit!"

Quinn asked how the safety system worked. The Old Finn gave a quick but detailed explanation of how the spring-loaded system worked in the event of cable break.

Running Fox, at the edge of the table, pushed the barrel of the carbine between Lubo and Nick. Tapping the folder with the tip, he asked, "What's down there? Must be something important."

113

It was the question Quinn hoped wouldn't surface. All eyes turned in his direction. He reached for the clipping taken from the Duluth newspaper, a four-column photo with a view of the empty cavern hewn from the granite at Level 27. He pulled several other articles from the scrapbook. "The project's called DANTE'S ORCHARD—one of the country's most vital research studies. Beyond that, I've been told nothing."

"Danti?" echoed Nick. "I knew a Danti from Virginia—Eye-talian for sure!"

"He was Nick," answered Running Fox quietly. "Lived in the twelfth century. Wrote a work called the Divine Comedy . . . about a trip through Purgatory, Hell and ended up getting a peek at Heaven and his old girl friend. Most the time was spent looking at the dark side of hell however . . . ."

"About hell eh—and he calls it a comedy?" Nick snorted.

Lubokowski pushed back, huge hand drumming on the sketch. "Don't mean to be disrespectful, Colonel, but you're asking us to take on the Russians, our own National Guard and some damn heavy weather. It doesn't look to me like you could drive a lubricated ten penny nail through that ring of tanks."

"I can shed some light here—a reason to take on the Russians," said LaMont hoarsely from outside the circle. He asked Karin to wheel him closer to the table. His hand reached out for the clippings. "If you read all of these you'd find the project referred to as a search for a missing part of a big puzzle. Different articles and each one has a different name for the part. Muon, monopole, neutrino, proton, quark and cygnet. Find these and you could possibly replace fissionable material as the energy source of the future. These particles with many names, are believed to pass through mass—such as earth—at the speed of light. So, locate your laboratory underground, create banks of cells much like old fashioned lead photographic plates and try to trap them in flight. If they exist—no more dependence on oil!"

Quinn suppressed a smile. The geophysicist was giving his men the straight facts on the cover project.

"Jahah, and maybe we gots something else to worry about."

Nick looked at his grandson, hand to his forehead.

114

"I just remembers, we could have friends and kin down there!" He turned to LaMont. "Colonel here said there are thirty-eight people in the mine. They can't all be University?"

LaMont nodded.

"Eighteen are *Boomers*—outsiders—like myself. Twenty-two are from around the area. The list is taped to that computer." He pointed beyond Quinn who turned to retrieve it.

LaMont continued, "Scientists, software and technical people are highlighted in yellow, local support staff in green. They took over the lab at shift change so everybody's a prisoner. Night shift should have been home hours ago so phones are probably ringing all over town."

"Damn! muttered Quinn. "Another problem!" Unfolding the paper, he scanned the list and began reading. Many had Finnish names, one a familiar Indian name. He looked around the small group and knew that all the men but Lubo had friends or relatives under Russian guns. Running Fox, shook his head slightly as if to deny the knowledge his sister was one of the captives.

Nick ground a balled fist into the open palm of the other. "Bastards, dirty bastards," he muttered. "No wonder she didn't answer the phone . . . ."

"Anyone down there besides Cousin Toiv in the machine shop?" asked young Saatela. Startled, Nick looked up, his face dark.

"Mrs. Pavo's down there. Day shift cook. She's a friend." He hesitated, "a very good friend . . . ."

Wally looked up at Quinn and back to his grandfather. "We've been out of touch, Grandpa. I didn't know . . ."

Quinn came back to the edge of the table. "It's time," he said. First we've got to find out where the tanks are, how the communications work and where that missing APC with the antenna is stowed away. If anybody has ideas on getting something out of the SPETSNAZ in the other room . . . let me know. Now, gather round, I'll make perimeter probing assignments . . . ."

LaMont nodded to Karin who pulled him back as the men stared down at the rough maps.

"I need someone familiar with the area to reconnoiter the backside of the ring. Terrain won't allow tank movement in the northwest quadrant."

Running Fox, leaning over the table, put his finger on the area. "I'll take that and the SPETSNAZ."

"And the SPETSNAZ?" echoed Quinn.

Jimmie nodded. "I'll circle the area, work my way around and pick up some things I'll need. When I get back, the SPETSNAZ will tell you whatever you need to know."

Quinn remembered the young man's SEAL background and nodded in agreement. "Wally, you and Lubo operate as a team. Come up from the south. Don't go in the road the tanks came in on but get as close to them as you can. Try to determine how far apart they are— what means of communications are being used. Keep solid objects between you and the tanks whenever possible. One APC out there is GSR equipped—ground-surveillance radar. There could be some fixed-position radar as well. *All* tanks have searchlights. Weather's in our favor for this part of the job and Karin has the sheets ready."

Quinn looked at the glum-faced Nick Huhto, his thoughts still apparently on the Finnish cook.

"You and I, Nick, we'll work our way up from the east, try to come in past the Alaska shaft using the main buildings and Head Frame to block out the radar. We've got to locate Captain Holmes of the Guard and pinpoint security around the Head Frame." His eyes swept around the table. "Avoid direct contact. Don't take unnecessary chances. And remember, Guardsman are as likely to get shot by the SPETSNAZ as you are. Everyone understand that?" He looked at each of his men.

"One last thing. Anytime you can pick up weapons or ammunition of any sort, do it! We need to scrounge. Check your watches." Quinn looked at his. "It's 0210. Wally, Lubo, give yourself ninety minutes. Jimmie, you've got two hours. And Nick, we'll operate on a two-hour time frame. We meet back here but make sure this place is secure before you move in."

He turned to Karin, standing in the doorway. She held white sheets, cut for head holes and pillowcases slit opened and pinned to the sheets for hoods.

"If the SPETSNAZ return, you and the doctor are on your own. Put the small lamp with the green shade in the porch window. On, it means there's a problem. Off, the house is approachable."

Karin nodded, handing out the homemade winter camouflage.

116

In silence, the men pulled the sheets over bulky winter clothing.

Quinn slipped on his thermal gloves. "One last thing, it's cold out there. Keep something on your heads at all times, gloves on and feet dry."

All the men smiled at the advice but said nothing.

Quinn gave his ragtag warriors an abbreviated salute. "We meet at the Rover in five minutes. I've got some last minute communicating to do with Washington and Doctor LaMont."

As they filed out. Karin held each man's hand briefly. Doctor LaMont held his emaciated fingers up in a partial salute. Quinn activated the radio.

"We're about to head up to the mine," Quinn reported. "Three mini-patrols coming from three directions. We'll try to determine where the tanks are, how they're communicating and what the defensive positioning is all about. The mission is set for go at two-twenty our time. Patrols are due back in two hours and we'll brief you. Mel, how goes the satellite search?"

"Two primary targets, Colonel. Both are Russian low-pass polar orbiters. General Henderson can tell you about the take-out plans. Just to confirm, it looks Russian all the way. If you can get the SPETSNAZ to talk at your end, it might be worth the effort. Here's the General."

"Caleb, the Russians have activated the Hot Line, declaring that the radar at Krasnoyarsk goes on in two hours for *test purposes only*. Orders are coming out of Moscow faster than the boys at Meade can decode and process. Even the Israelis have got their ears on. If this keeps up, the entire world intelligence community will be alerted. The President's worried this thing's escalating too rapidly and it's time for us to use the Hot Line." Henderson paused, "The President's authorized LaMont to tell you the true nature of the project. This is for your ears only—understand?"

"Understand Jocko. My men are waiting General. Look for a call back in two hours."

"Good luck Caleb."

"Thanks Jocko but first—a weather check at this end."

"Air Force meteorologists predict the storm will break about noon tomorrow—at the earliest, followed by a high pressure cell bringing intense cold to the region—twenty, possibly thirty degrees below

117

zero—wind chill factor to minus sixty degrees. You also asked for a re-cap on the number of Guardsmen and so-called regulars. We have a head count of 220 for the Guard—at least twenty-four for the regulars and possibly we're one or two shy on that score."

Quinn's head turned from side-to-side in frustration. "And I won't see any back-up for another eight hours? Hartlett could have his information out of the computers and in Moscow within the next four!"

"Do you want a sortie of all-weather fighters up there? We've managed to get a few top guns and wild weasels on the deck at Minneapolis. The plows can't keep up with the snow and drifts. Just one runway is being kept open. Flying time to Tower-Soudan about fourteen minutes. that's the best I can at the moment."

Quinn let a deep breath of air escape, his mind racing. "All the firepower in the world won't help, General, but I tell you what might. Send up the planes and have them arrive at 0310 hours our time. My men will be on the perimeter. The sudden appearance of a thundering herd might be the confusion factor we'll need. It'll telegraph to Hartlett we're on to him. He might make a mistake. Have the Air Force make as many low pass runs as time and fuel permit—but *no* bombing!"

"Air support—0310—your time Caleb. You've got it!"

Quinn snapped off the set swinging around to face Karin and LaMont. He briefed them quickly then rose.

"You want me to fill you in on the ORCHARD before you go?" LaMont looked up at Quinn, bony hands trembling.

"It'll have to keep," said Quinn looking at his watch.

Seeing the small disk on the edge of the table, he reached for it. Then, knowing he had no way to turn it on or off, he placed it back on the table and handed his Luger to Karin. "Use it if you have to." As he headed for the open stairway, Karin reached out, pulling Quinn into her arms.

"Take care, Caleb. I'm scared to death of what might happen to you and the others." He put his arms around her waist. Rising on her toes, she pulled Quinn's head down, her lips pressing hard against his, then released her grip and blinked back the welling tears.

Quinn pulled the makeshift hood over his watch cap, turned and descended the narrow stairwell into the dark morning and blowing snow.

118

Gathered around the Land Rover, the men were already stamping their feet against the cold. Quinn, sorting through his hunting supplies, found an extra compass for Wally. A compass mounted on the windshield was pulled off and given to Running Fox. Nick had his own tucked in the pocket of his Mackinaw. Quinn took the radar detector under the dash throwing it in a small bag with a 12-volt battery. His head brushed the top of the cab and the compound bow suspended by bungee cords. On sudden impulse, he unsnapped the cord, releasing the weapon. Jamming the bound bunch of arrows into the cloth case, he asked Saatela to retrieve a roll of heavy-duty monofilament from the back of the four-wheeler. He uncased his 30-06 scoped hunting rifle, taking two boxes of shells from under the sleeping bags. His hand touched the frozen partridge and for one brief moment, the vision of a candle lit dinner with Karin, complete with Caleb Specials, red wine and good food flickered through his mind. It seemed a hundred years had passed . . . .

Running Fox spoke to his brother-in-law in Ojibway. Wally grunted and pulled blasting caps out of the wooden dynamite box, handing them to his *niita.*

Quinn noted the exchange.

"Avoid gunfire at all costs. This is a reconnaissance—pure and simple."

Wind had built up drifts around the Rover. Visibility was limited to twenty yards. Dark pines swayed above them.

Lubokowski sucked in his breath. "Jeez, just open your mouth and you know how many cavities need fixing."

Wally and Lubo turned south, Running Fox north. In minutes, they were out of sight, only a deep cut in the snow indicating direction of travel.

Nick looked up at Quinn. "I could find the Alaska in my sleep. I leads!" Without waiting for a reply, he headed northeast. Quinn followed the small man whose white sheet dragged over the icy snow— the vintage *Soumi* tucked under his arm. They moved in silence. Snow crystals peppered Quinn's face, the wind carried his breath away, the cold brought tears to his eyes.

A seventy-year old man going off to war in one direction, his grandson and Lubo another. An Ojibway heading solo into enemy territory and a retired Colonel is back on the payroll up to his ass in miserable. Quinn snorted and wiped at his nose with the back of his thermal glove.

He kept his eyes on Nick's back as they worked past dark houses. Then the last of the private residences faded behind them. Entering a ravine. Nick stopped and turned. "Longer but safer. We can stay out of this radar you're talking about. When we gets to the Alaska Shaft opening, we'll have rocks and trees for cover. Near the Head Frame, once the ground levels out, we're either out in the open or we bellies in."

Quinn agreed. Drifts and bushes slowed their progress but they slogged forward, wind whistling over their head.

DANTE'S ORCHARD has to be something other than a search for errant particles of matter—what else can there be worth this much angst? thought Quinn . . . .

*   *   *

When the President re-entered the Situation Room, he was in a room transformed once again.

Cots, neatly made up, lined one wall of the long room. On one, a sleeping form snored peacefully. The nimbus of white hair nestled in the pillow told him it was Professor Chen.

Hart rose as did the Secretary's of State and Defense, Yoder and the Joint Chiefs of Staff. The presence of the entire group, a signal to the President that the situation had intensified during his two-hour absence. "Consensus is growing rapidly, Mr. President, that the Russians mean to follow this with military action. The signs are all there—radar at Krasnoyarsk up and operational, communications with Mir and Solyut space stations signal preparation for increased activity. General Benedetto, speaking for the entire military, recommends you implement the Doomsday Plane or we all move to Berryville. The plane can be here in 90 minutes. The back-up unit is having some new equipment added in Texas."

Incredulous at the suggestion of utilizing either the special command post aircraft or Mount Weather, 48 miles away at Berryville, Virginia, the President shook his head. "We've talked about that, Tom.

120

You can't start the engines on those things but that the press and anyone who's waiting for Armageddon comes to top alert. As for Berryville, you all know my thoughts on being run to ground . . . ."

Hart motioned toward the far end of the room.

Yoder, surrounded by the ranking officers of the U.S. Military, studied the incoming reports. "The military wants, Mr. President, at the very least—to keep up with the Russians. They escalate—we escalate. If the button gets pushed, we'll have a fair chance at retaliation." Hart waited for an answer.

Looking long and hard into the eyes of his National Security Adviser and then toward the group at the end of the conference room, the President shook his head. "No, Tom, not until I've had a chance to hear for myself what's going on!" Signaling with his hand, he motioned for Hart to awaken the sleeping head of the Defense Analysis Institute. Yoder, seeing the President take his place, motioned the military officers to take theirs.

Chen rubbed sleep from his eyes as the President called the meeting to order.

"I don't think there's a man in this room who won't agree that bringing one or both Doomsday aircraft to Washington under these conditions could push escalation out of control. It's never been clear to me how the KGB boys do it, but I believe they have watchers on the ground in Ohio and Texas who do nothing but keep an eye on those aircraft. When they hiccup or fart Moscow knows. If we can monitor the Blue Train under the Kremlin then, damnit, they can read our moves as well!"

Reaching for a silver carafe, the President poured a cup of coffee and looked again at the somber, drawn faces staring back. "Starting with you General Henderson and working around the table in clockwise order, give me a three-minute summary of the situation in your sector and a one-minute outline of action you'd propose . . . ."

On his belly on the tomb-like rock, the cold working through his down jacket, Quinn peered into the dark hole. Nick, beside him, pointed with the barrel of his vintage Finnish automatic. "This shaft still sucks air. Along with another shaft on the north side, they ventilate the entire mine with outside air."

"How far to the first building, Nick?"

The old Finn stared into the wind. "One hundred yards to the Head Frame and Crusher Building. Engine House is forty yards beyond that—Dry House between the two. Auxiliary buildings are scattered around."

"In a few minutes Nick, all hell should break loose. If we can cover the first hundred yards while the jets are screaming in we maybe could avoid the radar."

"Let's stay on our bellies, Colonel, and plow snow with our noses. But when your planes get here—we ups and runs like hell! I'm still your compass, I go first."

Quinn pushed the bags containing his supplies into the crevice, covering them with snow. Without waiting for his reply, Nick, automatic notched in the crook of his arms, slithered off the rock into the night. Quinn crawled after, inches from the Finn's hunting boots.

Snow piled up as the two men edged forward. Every few yards, they had to pause to sweep the dry, white powder from their field of vision. The wind, an incessant whine above their heads. Five minutes and forty yards closer, Quinn grabbed at Nick's booted foot. "Coming up along side. When the jets get here, I'll take the point. High Nick, I hear something at altitude."

"Jahah!" the Finn grunted. "Ain't gonna do us no good up there, Colonel."

"That's the Wild Weasel, Nick. He's got the eyes and ears, tells the all-weather bombers where to go and what to do."

Quinn's words were drowned out by a thundering roar shaking the frozen ground. The first blast of sound was quickly followed by a second. He struggled to his feet, pulling Nick with him. "Right on the money! Go, Nick, Go!"

Ears ringing, Quinn raced forward. The spidery shape of the Head Frame loomed ahead. He hoped Nick was close on his heels when he rolled to the ground and was relieved when Nick, panting, tucked in beside him. In the distance, they could hear two jets screaming for altitude. In seconds, only the wind filled their ears.

It was Nick who heard it first. He tapped Quinn's shoulder. "Twenty yards Colonel. Diesel running. Must be a tank guarding the entrance to the Head Frame and the cages."

122

Quinn was barely able to make out the dim outline of the white tank. Another few seconds and they would run under its gun barrel.

Suddenly, they heard muffled voices. Quinn patted Nick on the back. "That got their attention. Second time around we skirt the tank and head for the Dry House. Ready?"

"For sure!"

Head raised, elbows supporting him, Quinn tried to guess how long it would take the jets to be back. His thoughts were interrupted by the second, shattering blast of sound as a dark object appeared momentarily overhead, orange flame present for an instant. On his feet before the second jet screamed by, Quinn ran as fast as the thigh-high snow would allow. Suddenly, he fell, tripped up by something just off the ground but below the snow line. Nick, a step behind, fell on top of him.

"Damnit!" The Finn rolled off his back and came up snorting snow.

Feeling at his knees, Quinn found the object, pulling it up. He looked back in the direction of the tank, now behind them. In the darkness, he smiled to himself. "Nick! you OK?"

"Jahah, but what happened?"

"We got lucky. They're using landlines to hook the tanks together for communications. We follow this and if it doesn't lead to another tank, we'll find a surface command post." The muffled voice of the tank crew behind them was replaced by a voice ahead. On hands and knees, Quinn followed the wire, pulling it through the soft snow. The voices became more distant. Ahead, he could barely make out the dull glow of building lights. Nick, following in the trough behind, grabbed at his foot. Quinn stopped, waiting for the old Finn to move up beside him. He needed a breathing spell and Nick's advice.

"Dry House Colonel. That's a smart place for headquarters—big enough and heated."

Quinn tugged at Nick's white sheet. The Finn worked his way closer.

"Fighting the Russians, Nick. Tanks. Got any good ideas?"

"Same weather as this, maybe worse, we nails more tanks than you could shake a big stick at."

Sliding back, Quinn got his nose almost next to Nick's. "How Nick? What worked?"

"Molotov Cocktails and ice—leads a horse to water in the wintertime and the somofabitch'll freeze his balls."

Quinn glanced toward the light. Conserving his energy, the old Finn lowered himself below the snow line. "Best trick, with the Hunters—Jaegers we called 'em in the old country—near Lake Ladoga—we sets up an ambush. I was in the ghost patrol, a snot-nosed kid. We went to the Russky tank lines one night dressed just like this . . . fired a few shots . . . when they moved toward us, we drops back, keeps the firing up—just barely—to keeps their heads low in the tanks. When they gets to the lake, by gully—another target—snow fort on the ice. A few crazy Finns like me ran around, pretends to be hundreds instead of just six men—and they fell for it. Russky commanders thought they had us out in the open for a change and moved out. Fifty feets from shore, our explosives man press the plunger. Balloons made of horse guts tied to waxed explosives. They be floated up under the ice in a string. CABOOM! Tanks sank up to their turrets—just where we wants'em. We went in guns firing. Their infantry die or retreat. We strips the tanks for guns."

Quinn got up on his hands and knees and pushed on, following the wire, hoping it went where he needed to go, Nick's Talvisota story finding a place in his memory. Ten yards from the light, he stopped, dropping onto his stomach waiting for Nick to edge in beside him.

"If the one guy I want is going to be anywhere, that's the place, Nick." He dropped the wire, satisfied it led into the Dry House. "But I gotta be sure."

"Only a few yards Colonel, I can get up there."

Quinn shook his head. "I need you here. You're the only source of information I've got."

A sharp blast of gunfire in the distance pierced the wind. A long burst, followed by two short, then silence. Quinn knew it was a tank-mounted M60. An instant later, a battering ram of sound assaulted their senses as two jets, this time side-by-side dropped low over the mine site. A door ahead of them burst open and a man stood framed in a halo of light, head tilted back, trying to track the climbing jets.

"Close the damn door!" A voice bellowed behind him, followed by the sound of a door slamming shut. For an instant, the man's outline disappeared in the blowing snow. When it re-appeared, he was less than ten feet away, coatless, back turned, urinating in the snow.

Quinn whispered in Nick's ear. The old Finn nodded in response.

Rising out of the snow, Quinn covered the short distance in two strides. "Let it hang soldier. Don't bother tucking it back in." He jammed hard in the man's back with the barrel of the 30.06.

"What the!. . ."

"Shut your mouth and raise your hands," hissed Quinn as the man's hands flew up.

"Turn, walk ten feet and drop to your knees." He pushed the barrel into the man's mid-section. "Now Drop!"

Obediently, the man fell. Nick was nose-to-nose with the wide-eyed soldier.

Positioning himself so he could see the door, Quinn dropped down.

"You Guard or Regular?" He put the ice-cold barrel against the man's neck.

"Shit man! I'm Guard! What'n hell is going on?   I'm starting to freeze out here!"

"Quick, Nick. Run this guy through his paces. Is he native or ringer?"

The old Finn pulled the young man toward him, clutching his shirt at the collar.

"Name?"

"Jutkaninen."

"Home?"

"Virginia."

"What church you go to?"

The man stammered, teeth starting to chatter from cold and fear.

"I. . I don't go to church."

"Damn young . . ." Nick pulled the man closer.

"Your parents then . . . what church?"

"Fir . . First Cov-Covenant Lutheran."

"Pastor?"

He shook the guard. "P-p-pastor Lukoven!"

125

"Name of the high school football team in Virginia?"

"Blu-Blue Devils!"

Quinn pushed the young soldier down in the snow as the door to the Dry House swung open.

"Corporal Jutkaninen! Report back here immediately!"

Quinn murmured in the man's ear. "Who's that?"

"Sergeant Wright. F.F.First Air Cav. Umpire—regular army."

Sticking his head up over the snow line, Quinn could tell the man wasn't dressed to conduct an outside search. He turned and closed the door. Quinn guessed he had sixty seconds before the SPETSNAZ returned and followed the foot prints.

"How many other regulars in the room?"

"Ju . . . ju . . . just one. A Captain Zeller."

"Where in the room was he when you came out? Quick, Damnit!"

"Center, the radio console."

"Holmes, Captain. He in the room?"

"Yes. Cross from the other Ca . . . Ca . . . Captain."

The young guard's teeth began to chatter but no intelligible sounds came forth.

The door to the Dry House opened and a bulky form stepped out. The door shut behind him.

Quinn put his mouth to Jutkaninen's ear, his body pinning the young Guardsman in the snow. "Lay face down like you passed out. You understand me?"

Shivering, body convulsing, the soldier's response was almost muffled.

He pressed the man's head into the snow.

"Your automatic Nick, I'll need it. Fall back. Get covered with snow."

Quinn rolled through the snow away from the prone Guardsman, pulling the hood low over his face with one hand, pushing the Finnish machine gun under snow with the other.

Head down, he waited on his knees.

A jet screamed in from the South. He repressed an urge to look up. When the second aircraft thundered in at right angles seconds later, he lifted his head slowly. The SPETSNAZ was on his knees behind the

Guardsman, M16 in hand. Sound reverberating everywhere, Quinn gripped the heavy Finnish weapon by the barrel and swung it like he was driving tent stakes. The topside of the wooden stock caught the Russian between the shoulder blades. He collapsed with a groan on top of the Guardsman.

Huhta popped out of the snow.

"Cut his laces, Nick. Tie his thumbs behind his back and his feet together. He stays out here until we buy visiting rights to the Dry Room!"

Pushing the Russian off the younger man, Quinn yanked the Guardsman to his feet. "Get the snow off and get ready to go back in ahead of me. In the door, you face Zeller instantly, wherever he is, then drop flat on your face. You understand me, son?"

Tears freezing on his face, the young man nodded dumbly.

Quinn patted him on the shoulder, steering him toward the doorway.

"You can zip-er up soldier. You did fine. Just fine."

Hands shaking, the young man stumbled forward, unable to manipulate the zipper on his urine-soaked fatigues.

Outside the door, Quinn stopped the soldier while he slipped the safety on the rifle and listened for jets. "Now!" He pushed the soldier with his hand.

Stumbling, the Guardsman caught himself and entered the room. Momentarily confused, he turned to his right.

Quinn stepped into the room. Aware of others, he kept his eyes on the back of the young Guardsman's head. Jutkaninen's body pitched forward, arms shooting out to break his fall. Quinn saw the hand reach for the weapon on the desk in front of the green radio console and he caught a glimpse of what he was looking for—Captain's bars and heard the sound he prayed for. Men dove for the floor on either side of the captain. Eyes on the bulge under the sheet, the Russian had the automatic in his grip when the .30.06 hollow-nosed round caught him in the chest, lifting him off his chair and splaying him against the wall. The windowpanes rattled in their frames from the Air Force bombers overhead.

Quinn quickly pushed back the white hood.

Ernie Holmes, first man off the deck, sidearm drawn, recognized his former commander emerging from under the ice caked sheet. Adrenaline pumping, Quinn nodded silently but kept his rifle at hip level, back to the door. Jutkaninen lay on the floor, quivering.

Hearing a thump behind him, Quinn, eyes scanning the room, pulled the door wide open as Nick, using his rump, edged in backward, dragging the unconscious Russian.

"Enlisted men, Ernie? Are they yours?" Holmes nodded. Two privates came up off the floor, eyes wide. The redheaded Guard officer stared at Quinn, his gaze going back to the dead captain—the regular army sergeant lying bound on the floor.

Quinn understood the captain's conflict. "You're looking at Russians, Ernie. There isn't a regular army trooper in the bunch. This maneuver was pulled off by the Russians to capture whatever it is the computers below are working on."

Holmes still seemed uncertain.

"Nick, you got a knife?"

A curved-blade knife materialized out of Nick's pocket—a larger version of Karin's puka.

Unzipping the battle jacket and the top of the woolen sweater, Quinn used the razor-sharp knife to cut into the thermal underwear, exposing the man's upper left chest. He quickly found the outline of the disk. One thin cut and the coin-like object was out. He wiped it clean on the bed sheet and handed it to his former reconnaissance leader. The Russian trooper remained unconscious. Quinn patted the thermal underwear back over the incision.

"You'll find one in the dead man's chest as well. In addition to being a type of dog tag, it paints a signature on S and J-Band Radar. This is how Snowtop down below, keeps track of his men." Handing the puka back to Nick, he nodded toward the dead man. "After you get his disk out, Nick, roll the guy over. Let's see what his back looks like."

Dropping to his knees, Quinn flipped the live Russian over, pulling up the sweater and thermal underwear. Fingers of raised welts crossed the man's backbone. In minutes, Nick had the dead man's transponder and confirmed that his back, too, was a mass of scar tissue.

Holmes squatting next to his former commander stared at the men's backs and then at the disk. The green eyes blinked.

"They're all like this?"

Quinn nodded. "They've picked up two in Washington, a man and woman. Same thing. Imbedded I.D.'s doubling as position locator. If we had a way to read these, we'd find their M.O.S . . . same as the Military Occupational Specialty we have in the Corps. Probably would learn they served in Afghanistan before being assigned to this mission . . W might even come from the same general area in the Ukraine for some strange reason."

"The jets, Colonel. What's with them?"

"Ours, Ernie. A buzz job diversion to our penetration of the tanks. They're armed but under orders not to attack unless cleared by me. You've got 220 National Guardsmen exposed with Russians mixed in. I need to know where they are. Another risk are the Stingers – no question in my mind these guys know how to use them!"

Holmes rose to his feet. "Two in the Engine Room behind us. Three in each tank on the line, two in the Command APC. There must be two more down in the mine."

Making rapid calculations, Quinn looked up, "That makes a total of twenty who came from the outside. One is dead, two prisoners and two appear to be infiltrators who worked their way into the project as moles. We've got a minimum of twenty-eight to eliminate or neutralize—more than I figured!"

"How many men do you have with you Colonel?" Holmes looked at the old Finn. Nick apparently read Holmes' mind.

"Jahah, son! You're looking at almost one half of Quinn's army. There's only three more out there."

"What's wrong Ernie?" asked Quinn, catching the dismay.

"You might have lost someone, Colonel. Just before the jets came screaming in, our tank on the southern perimeter identified two targets on radar—I heard a regular army NCO give orders to open fire."

Holmes realized he was the object of intense scrutiny by the old Finn.

"Young Saatela, Nick's grandson Ernie and Lubo . . . two volunteers . . they had the southern sector—Jimmie Running Fox the northwest quadrant."

129

Quinn had a sick feeling. "Your commander, Ernie. Where's he located?"

"In the APC, and I'm not sure exactly where that is. I was told to maintain my position here and take orders from the command track."

"That command track Ernie, it's important that it be located. That's probably the sending source for the data coming up from below. Our job is to stop it from happening—and keep the casualties to a minimum. Washington is looking at a Russian military build-up that seems to coincide with this effort."

Holmes turned to his two enlisted men. "Meet Colonel Quinn, my commander in 'Nam."

"And now I'm the field commander for this job," responded Quinn.

Holmes smiled and saluted. "Priorities, Colonel? What needs to be done?"

"Straight forward, Ernie. Stop Hartlett dead or keep him from sending the data. You can take on one hell of a responsibility by getting your guard out of here. I won't authorize any bombing until I know we've somehow isolated the Russians. When the weather breaks, a cold-weather Marine brigade with two companies of Norwegian ski troops will be airlifted from Camp Ripley. Your men will be integrated with them. The other concern is the local citizenry down below. Thirty-eight people in the mine could die if things don't break right."

"Wow! Any other orders, Colonel?"

"There must be a link from the radar down into the mine and Hartlett has a picture of what's going on. Shut it down, and I guarantee he'll squirm just a little more. Right now, he probably thinks these disks represent two live soldiers he can count on. One last thing—write down the frequencies you're using. Maybe I can patch in from the radio back in Soudan."

Holmes took notes, then pointed toward the corner. "We've got a RAT Rig Colonel—collapsible antenna and it's patchable to satellites."

"Take it. We need all the links we can get a hold of. We also need hand-held radios and grenades. What've you got?"

The Guard Captain ordered his men to round up what was available. Quinn shoved six grenades in his pocket and ordered Nick to take what he could carry. He selected one of the two remaining field radios, slinging it on one shoulder. He traded his hunting rifle for a tanker's grease gun and four full clips. Just before leaving, he turned to his reconnaissance leader.

"Confirmation code, Ernie. Let's dust off the last one used at Khe Sahn."

Puzzled, Ernie scratched his head then broke into the grin Quinn remembered so well. "You've got it Colonel!"

Pulling the hood up, Quinn eyed the two Russians. "Find a home for the dead one and make damn sure the live one stays secure until this is over."

"Yes Sir!" responded the Guard Captain.

About to slip the transponder disks into his slash pocket, he remembered that they must still be sending off a signal. He put the twin disks on the window frame over the doorjamb—out of sight. *Keep somebody guessing* he thought to himself.

*   *   *

Quinn and Nick talked over the strategy of the return trip. Knowing each tank had a Russian on board discouraged short cuts. Quinn took the lead, on his belly, through snow that had deepened by several inches. Wind raged from the west, causing their white parkas to billow and snap in the wind like dozens of whips being cracked. The cold seemed more intense coming out than going in. Concerned about Nick, Quinn paused, waiting for the old man to close with him.

"Break time, Nick." he paused, "I hope Wally and Lubo made it back."

"Yah for sure, Colonel. Wally's got a wife, a son and another on the way." Nick paused, "I've been thinking, Colonel, plowing through the snow back there. I lived a long life, ain't seen everything but seen a lot. Wally now, that boy's got a piece to go. Just a dumb Finn though—marrying a Catholic girl—with Indian blood. Not too smart."

Ready to slip between the rocks making up the Alaska opening, Quinn stayed low but Nick stood up and a second later crumpled into Quinn's arms as a rifle shot shattered the night.

Shot-or shots, Quinn couldn't tell as he dove in the snow-covered rim of the boarded shaft. He pulled the old man onto his lap, a dark circle blossoming on the left side of Nick's temple.   Using the bed

131

sheet, he tried to staunch the flow and determine if Nick was dead—or dying from the shot to the head.

Had to have been a night scope . . . why'd the little guy have to stand up?

Pulling the flashlight from his pocket, he examined the wound. Blood came in even spurts. He pressed a clean part of the hood onto the glistening flow then used the light to check Nick's pupil. Out of the howling wind, eddying snowflakes dropped onto the wound and quickly melted into darkening mass.

Quinn realized Nick was his one and only key to getting into the mine. Now the feisty veteran of another war against the Russians lay moribund in his arms.

He lifted Nick's head, cradling the miner in his arms for warmth. Nick's eyelid flickered then snapped open. The blue eye was round and centered. Suddenly the other eye blinked open. Nick's gaze was steady.

"What in hell happened?" he muttered hoarsely.

Relieved, Quinn suppressed a smile, "Must've been infra-red Nick. Maybe you got grazed or hit by a rock sliver—your gonna be OK."

Quinn helped the older man to a sitting position.

Nick shook his head, like a Labrador coming out of water. "Oh ya—remembers now. We were heading home. What were we talking about?"

Quinn's grin was hidden in the dark recess of his makeshift hood as he secured a strip around the old man's head. "Intelligence Nick."

"Holy Mother, what happened?" Quinn knelt beside the heavy breathing Lubokowski.

"Lubo saved my life, Colonel."

His head jerked up to see Wally, eyes wide open, gesture with his hands.

"We came up right under the nose of a tank," he said. "Lubo found the interconnect wires and we moved from tank to tank until we got to Number Four. Figured we had the information. It was the tank with radar that almost got us. We crawled until we got into thick pines and Lubo ran smack into two spike bucks bedded down. They jumped up

132

and in seconds the trees were snapping over our heads like matchsticks from M60 fire. Lubo took a hit in the shoulder when he reached up to pull me down. The deer just disintegrated . . .

Hands on his knees, Quinn looked at LaMont's empty wheel chair, then to Karin.

"He's on his bed, very sick and very discouraged. If we don't get him some kind of help, Caleb, I think he's going to die."

"Did something happen?"

Karin shook her head up and down, eyes starting to mist over.

"Hartlett killed his friend Bochert—right before his eyes. Maria is next unless he tells that son-of-a-bitch whether the glitch was put there on purpose. LaMont told me it was and could be removed in minutes and then he slipped into a sweaty coma like we've seen before. I put him to bed before he fell out of the wheelchair."

"Hartlett's coming back on the tube?"

Karin glanced at her watch and nodded. "Ten minutes."

"The prisoner?"

"In the root cellar. Jimmie came in ten minutes ago with a wet gunnysack. Jerked the prisoner up and marched him to the basement. Said it would go faster if I came down. I asked why. He said it had to do with masculine pride. I turned him down. I've seen enough suffering."

"How do you get to the root cellar?"

Karin pointed toward the kitchen. "Small back stairway."

Nick clumped up the stairs and into the room, Karin's hands flew to her mouth when she saw the blood. Wally pushed himself to his feet and stumbled to his grandfather, his arms wrapping around the old man.

Suddenly weary with the responsibility, Quinn's eyes closed and slid down, back to the wall, hands coming to his eyes.

Dropping to her knees, Karin put her arm around him. When he lowered his hands, the eyes were red-rimmed. He rose unsteadily to his feet.

"Gotta check on Jimmie. If you can find something for these guys to eat, go to it. They need something."

"Gonna put a big pot of chili on. Theodopolus was crazy for it. Two cases in the kitchen. I've got twelve cans open."

133

Quinn worked his way down the narrow stairs, the small flashlight flickering in the musty passageway. At the top of the stair leading from the ground floor to the cellar, he heard a low moan. Eight steps and he was on the dirt floor staring down at the Russian. A small bulb dangled from the cob webbed ceiling casting a pale light over the small room. The Russian, naked except for his bandaged foot, was bent over an old fashioned, brassbound chest, legs spread wide. A rope was looped around his neck and secured to a concrete block resting on the sand floor. There was enough slack between block and neck to allow the man's head to move. Fresh blood trickled from the wound on his back. At Quinn's appearance, the young Russian's head jerked up, eyes pleading. Behind him, striped to the waist, four black lines radiating at angles down his cheeks, Jimmie Running Fox, arms folded sat on an upturned box, face expressionless. Between his legs and directly underneath the bound man's exposed backside, was a five-gallon bucket. Something in it moved, splashing water on the prisoner's legs. The Russian's body twitched.

"You have questions, Colonel? Our man here has answers. Unfortunately, he seems to have forgotten English, but he's very talkative in his native tongue."

Running Fox reached in the bucket and lifted out a mottled-green snake pickerel behind the gill plates. The eel-like foot-long fish twisted in Jimmie's grip. Quinn realized what the exchange of blasting caps between Wally and his *niita* was about.

"*Ogan.*" said Jimmie, face expressionless.

He let it flop against the man's buttocks. The young soldier's head rolled and the same piteous moan broke from his lips. Grasping the tail with his other hand, Running Fox let the cold, mucous-covered body caress the man's still swollen testicles and then pushed the elongated, bony nose of the pickerel against the man's twitching anus. The Russian's eyes were those of a pleading puppy. Quinn felt like vomiting.

"Enough, Jimmie! He's probably over the edge."

Kneeling in front of the Russian, Quinn quickly cut his bonds with the K-Bar, pulling the Russian to his feet.

"Help him get dressed and tie him securely. We'll leave him down here."

134

Running Fox, unperturbed by Quinn's action, nodded.

Barely containing his disgust, Quinn stared at Running Fox.

"You learn this in Vietnam?"

"Closer to home—Sioux graciously introduced it to us Ojibwa several hundred years ago when they sent raiding parties into our camps looking for winter stores—became known as *nin kotagima* in our language . . . ."

Shaking his head, Quinn turned and made his way up from the root cellar, it's dank odor made even more malodorous by human excrement and the pungent smell of raw fear.

*   *   *

Quinn walked into the dimly lit bedroom, odor of rubbing alcohol heavy in the air. Lying fully clothed on the bed covers, the geophysicist's hands were clasped over his sunken chest. His eyes followed Quinn but his head didn't move. Blizzard force winds rattled the windows.

"Sorry about Bochert, Doctor LaMont."

LaMont's thumbs twitched but he said nothing.

"Karin told me about Protocol 019—you're possibly being on placebos—I've got Washington working on getting medicine up as quickly as the weather clears."

LaMont's head rocked from side-to-side. "Won't help Colonel. My body's a dead battery." He started to cough, spasms racked his wasted frame. Quinn fought the urge to call Karin.

Tears rolled down LaMont's face and he pointed toward the tissues. Balling a handful, Quinn placed them in the man's grasp. The siege mercifully short, LaMont wiped phlegm from his chin cupping the damp tissue in his hand. Taking a deep breath, he moved to push himself up. Quinn helped, putting pillows behind the man's back.. "Shouldn't lay down—fluids collect in the lungs," he said hoarsely. "And regarding the study Colonel—don't make any special effort on my behalf—it's far too important for those who still have a chance for even one person to change the outcome by switching programs and with Teddy . . ." LaMont's eyes glazed over, voice trailing away.

Turning, Quinn shut the bedroom door.

"You're here to learn the true nature of DANTE'S ORCHARD?" asked LaMont, dabbing his eyes dry.

"That, and keep you feeling as good as possible." said Quinn as he pulled a bentwood chair next to the bed. "Without your inside knowledge of the project, we don't have much chance of stopping Hartlett and his men."

Quinn pushed another pillow behind the ailing man to support his back.

"You impress me, Colonel. For a military man without any military, you've accomplished a lot with very little."

Quinn dismissed the compliment with a wave of his hand. "I haven't done squat! One Russian is dead, two are prisoners and we have a Guard captain trying to separate good guys from bad out there. Not a damned thing's been accomplished underground and time is running out."

LaMont looked at Quinn through watery eyes. "I know you don't have much time but I've got to say this . . . you picked up on my having AIDS and didn't bat an eye. And your friend, Karin, both of you, no sign of fear or revulsion—just acceptance."

Quinn smiled. "Karin told me she always wanted to be a nurse. Me, I once thought I wanted to be a doctor . . . ." He studied LaMont's thin face and had second thoughts telling how he fainted the first time he saw blood—his own as a Boy Scout—and vowed never again to blanch at the human condition—regardless of state.

LaMont pointed his finger at Quinn; "I think you would have made a good doctor," a smile creasing his face.

"*Maybe* a better healer than warrior, Doctor LaMont, but time is short—time to continue."

LaMont wiped at his lips with a fresh tissue and tilted his head back, eyes toward the ceiling. "My calculations tell me Hartlett will have his simultaneous run between eight and nine this morning . . . if to save their lives, I tell Kubedera to clear the so-called glitch."

Quinn edged closer to the bed. "When it's complete What'll he have? I know it's not a proton caught with it's pants down or a monopole mating with a muon . . . ."

Smiling, LaMont pointed over Quinn's shoulder in the direction of the computer room. "The big globe out there—that doesn't look like any surface you've seen before . . . . ?"

Quinn's bobbed in remembrance.

"It shows the twenty-odd tectonic plates making up the earth's mass—major and minor plates—the earth's outer shell as it were."

"There's some relationship between it and what's happening under our feet?" asked Quinn.

LaMont's head bobbed, smile fading. "When the entrained CRAY's, all four of them stop, Hartlett will have in one hand, a precision map showing what the world will look like *after* the polar shift takes place and those plates realign themselves with earth shattering force." He paused, wiping his forehead. "In the other hand, he will have a very accurate timetable of *when* it will occur . . . ."

Pushing back in his chair, sucking in a deep breath, Quinn was silent. His mind trying to grasp the enormity of such an event. Several moments passed before he spoke. "Poles shifting out of position—ice caps moving on a massive scale?" He gave a brief shake of his head.

"Antarctica first, Colonel, its loading mass has been approaching the critical point for several decades—accelerating in fact. And now you know why there are so many research bases down there—virtually none at the North Pole. You've heard of the *greenhouse effect*—man's contribution to world well-being?"

Quinn nodded. "Too much hair spray on the loose . . . some relationship to the supposed *ozone hole* at the South Pole . . . ?"

LaMont dabbed at his chin with the balled tissue. "Exactly—the overall global warm-up makes that much more water available for ice formation. In addition to your *hair spray* reference, the increased use of fossil fuels over the last two hundred years have caused Antarctica, for example, to have its mass increased by 300 cubic miles of ice each year for the last twenty. That's roughly the size and volume of Lake Erie! Add the compounding factor and you've got some idea of the magnitude involved. And now, consider the sudden impact of oil fields that have been burning for months in Kuwait. Millions of tons of solid particulate being injected daily into the atmosphere, slowly being moved by Coriolis effect around the globe—the implications of that alone are . . ." LaMont's head jerked and he began coughing again. A film of sweat glistened on his forehead.

"Maybe you should . . ." Quinn started to rise.

LaMont's hand came up. "Sit Colonel. You need to know." He continued. "Given this type of loading, Antarctica *must* shift—Newton's Law. It can be triggered by events having nothing to do with the loading itself. That's why the CRAY's are critical. For the first time, we have the ability to make the rapid calculations needed to arrive at a reasonable calendar of events. The acronym I referred to earlier—LMRP—is the LAND MASS RE-STRUCTURING PROFILE, the attendant Computer-Aided Design Map—a graphic analysis of what this globe will look like *when* the shift is over. Who ever has it knows with reasonable certainty which parts of the earth's surface will remain in place, which will disappear and where new land masses might be formed. To further complicate matters, the infusion of ice water in massive quantities into volcanic seams will cause not only accelerated plate movement due to the lubrication factor, but tsunami's of a scale never dreamed of. Ice cold water meeting tropical waters will cause meteorological phenomena never seen before—the world will be convulsed at sub-surface, surface and atmospheric levels simultaneously! Hundreds of millions of people will die within hours. Consider the cyclones that ravage a place like Bangladesh and then consider Florida, already at virtual sea level as is much of the East Coast—even a small tsunami would kill millions—try to imagine one that could be a wall of water three hundred feet high, propelled by energy that could push it to the eastern slope of the Rockies . . ."

Rocking back and forth on the chair, Quinn was silent, his gaze fixed on LaMont's deep-set eyes. Neither man spoke.

LaMont coughed. "But it's the Shamals colonel . . . ."

Quinn's eyes narrowed. "*Shamals* . . . ?"

"I mentioned the millions of tons of oil coated desert sands being injected into the atmosphere . . . ?"

Quinn nodded, "I remember."

"Last week in London—taking a break from the Quark Conference hosted by Cambridge University—the U of M's partner for the cover project—Teddy was covertly given a set of floppy disks that represents the first in-depth analysis of the massive movement of the oil-coated particulate throughout the earth's atmosphere. We've had a chance to

138

review that data—make some preliminary runs on the CRAY's and . . ." LaMont's voice faded. his eyes closing.

Moving closer to the bed, Quinn asked softly, "And . . . . ?"

LaMont's eyes opened slowly. "And the data is devastating—simply devastating." His eyes fixated on the ceiling as he forced himself to continue.

"Not only is the amount injected far greater than expected, but the duration time it is held aloft is exceeding everyone's first guess. The particulate is so fine, hammered into almost a powder-like consistency by the Vietnam-style carpet bombing of the B-52's, that it may stay suspended for years affecting the weather patterns and that . . . ."

Quinn finished the sentence, "will speed up the *greenhouse effect* which in turn will add more ice in Antarctica and we'll have your *Polar Shift* sooner than expected . . . ."

A sardonic laugh broke from LaMont's lips. "You are a quick study Colonel and partially right. The *greenhouse effect* will be slightly negated but eventually global warming will increase again."

Before Quinn could respond, LaMont raised his right hand as if to stop him. "There something else you should know."

"More bad news . . . ?" asked Quinn softly.

"Yes and No. *No* if scientific discovery is important—*yes* if that same discovery somehow relates to an impending polar shift. You're aware, Colonel, of the four fundamental forces of nature?" LaMont waited for Quinn's response.

Scratching his chin, Quinn pondered the question before answering. "Ah, gravity, electromagnetic, nuclear and . . ." A smile on his face, he shook his head. "One reason I'm not an astronaut, Doctor—wasn't that great in physics."

"Three out of four's a passing grade in my class," said LaMont. "Nuclear actually represents two forces—strong and weak. To get to the point, Teddy Theodopolus found the key to identifying the Fifth Force. The scientific community worldwide is in common agreement, a Fifth Force, in all likelihood, exists but no one has been able to identify its mechanics. Teddy's work for the ORCHARD and close involvement in the cover project—Proton and Double Beta Decay—led him to the conclusion the Fifth Force has been staring us in the face all along and we didn't recognize it. '*Eureka*,' as he called it, occurred when we watched the Olympics in Korea—in fact, he named his hypothesis, *Force Olympiad.*"

"What's the connection to the pole shift?" asked a puzzled Quinn.

Hacking once, drying his lips, LaMont's eyes grew animated. "A very direct connection—the moment we saw the games opened by skydivers forming the five-colored Olympic rings—interlocked, I saw only a cleverly executed symbol of unity through sports. That night, Teddy had a dream—in the manner of Kekule and Bohr."

"Kekule—Bohr?" asked Quinn, brow furrowing.

"Friedrich Kekule, brilliant scientist, dreamt a serpent was swallowing his own tail and out of that vision came the realization that the molecular structure of benzene was a closed carbon ring—a major breakthrough. Niels Bohr, Nobel winner, said a dream showed him the structure of the atom—and the key to the bomb."

Quinn nodded in silence. The ghostly face of Hole-In-The-Day passed through his mind—that of the Ojibwa shaman who had taught him as a teenager that dreams were a source of inner strength—knowledge and sometimes—salvation.

"Teddy, brilliant scientist that he was, saw in his dream, a new form of energy—short-lived but extremely powerful! As the jumpers slowly came together, into separate, then joined rings, he instantly theorized that the four existing forces, working in subtle, but perfect harmony with one another, are the primary constituents making up the Fifth Force—that they assemble and disassemble in rhythms or waves so massive no one was able to contemplate their true nature or source. From that point on, whenever we could, we used the CRAY's to confirm his theory—creating theoretical models, crunching massive numbers, discarding those without merit, saving those that did. It was his moment of divine genius—inspiration, if you will—that put the project on the track that led us to this week. Had he lived, he would be acknowledged as the discoverer of Force Olympiad—most likely trigger mechanism to start a Polar Shift!"

Quinn was as startled by the revelation of impending scientific discovery as the energy manifested by the ailing scientist. His eyes glowed with excitement. "All of Teddy's work is in the computers. What remained to be done, locked in his ...." LaMont's eyes closed momentarily, then blinked open.

140

"This discovery, if proven out, would be a significant event?" Quinn asked, already guessing at the answer.

"Proof is close at hand." Wiping his eyes, LaMont brightened again. "As to its significance—it would rank along with Newton's Law of Gravity, Einstein's Theory of Relativity, Fermi's work in the atomic field. There's some doggerel scratched on a toilet stall down at the University. "Roses red—violets blue. Find Fifth Force. Nobel for you."

"Who knows about his work besides you?" Quinn saw the smile fade from LaMont's face, thin hands clutch at one another. "Originally, two people—Teddy and myself. Now, it's still two people—you and me. Dr. Chen only knows that something important is about to spring out of the CRAY's . . ."

LaMont's gnarled, almost fleshless left-hand rose; "One last point—regarding proof—based on the models used as inputs. Should the computer run being processed at this moment indicate a pole shift occurring within the next hundred years, it will be proof positive that Force Olympiad is fact, not theory."

Already stunned that such an event could take place at all, the revelation it could take place within the foreseeable future left Quinn speechless. Lips pressed together, he shook his head from side-to-side.

The storm in the desert seems not to have ended," he thought to himself.

When he looked up, LaMont's eyes, alive and piercing, locked on his.

"You are so right Colonel Quinn. The disks given to Teddy without authorization of the Saudi government? They commissioned the study after it was determined the oil fires could not be put out as quickly as anticipated. They contain still another bleak message . . ."

"And that is . . ." Quinn rubbed his forehead.

Teddy studied this information given to him by a young Saudi he'd befriended at CalTech during his graduate days. The friend knew the data was critical but he also knew his government was not likely to release it for fear of being tainted by its dark message—that they had solicited our help to reclaim Kuwait and in effect—were part and parcel of the problem. The linkage of the Shamal Winds and Force Olympiad is truly devastating."

141

Quinn leaned back, his eyes closing, his right hand massaging his temples.

"You seemed perturbed Colonel . . . ."

For several moments, Quinn held his tongue. When he spoke, his words came slowly. "I've had more than one occasion to learn first hand about intelligence—where it comes from—what you do with it. I was on the line at Khe Sahn, a classic case of misplaced intelligence that almost cost us an entire marine division. Then there was the incident in Beirut—that our installation there was going to be attacked was a *given* but that INTELL didn't get to the right people in a timely manner. Now I hear this and I wonder . . ."

Hand coming down from his forehead, Quinn stared at the geophysicist in silence.

"Tell me about the linkage Doctor LaMont."

"Its very simple for being a complex meteorological phenomena . . . the fact that the particulate is oil coated yet capable of being held in atmospheric suspension creates a massive electrical grid girdling the planet earth. Teddy's first computer runs seem to prove out that the *greenhouse effect*—global warming is going to occur on an accelerated basis and . . . if a Fifth Force truly exists, the presence of this airborne friction plate is likely to be the . . . ."

Quinn looked at LaMont then realized he was searching for the right word.

"The armature Doctor LaMont . . . ."

"You amaze me Colonel. Precisely the right word . . . the world is wrapped in these filaments and under the right circumstances the Fifth Force energizes—the motor keyed and then—when the convulsions begin . . . Ampere's Law is proven again—on a planetary scale!

"And what could trigger the *convulsions*," queried Quinn.

"Ah—an interesting question. We have still another problem to contend with Colonel. You're familiar with *compounding problems*?"

Quinn responded with a smile; "Yes—we called that '*clusterfuck*' in Vietnam . . . ."

LaMont's face grimaced but he too smiled. "Not too elegant a phrase but most apt. In 1994 the Shoemaker-Levy incident occurred. You recall what happened there?"

142

Quinn pursed his lips for a moment then responded, "the *'string of pearls'* – asteroids or pieces of a broken comet slamming into Jupiter?"

"Good memory Colonel. We're still not one hundred per cent sure *which* they were but the important fact is that twenty or so objects, some as large as two kilometers, slammed into the planet. Huge plumes of gases and debris rose thousands of kilometers into the atmosphere—heavy traces of ionized magnesium have been detected by the Galileo, Ulysses and Voyager II. The Hubble Space Telescope tells us this debris, like that of Pinatubo in 1991 and Gulf War have found their way into *'armature-like'* filaments circling Jupiter and . . . ."

"Setting up the same problem on another planet in our system," answered Quinn.

"Precisely," muttered LaMont softly. "And, that is what led to Teddy's final conclusion." The doctor seemed to Quinn to sink deeper into his pillow. A soft glow of light cast dark shadows under his eyes.

"Go on," urged Quinn.

LaMont coughed once, then continued: "Jupiter, being much larger than earth will be the bigger *'motor'* so to speak. When it *'kick starts'* so to speak, the dynamics of entire universe will change—the existing harmony between planetary electromagnetic fields will abruptly change. A *'fifth force'* will energize and . . . and . . ." LaMont's hand came to his mouth, a tissue covering his lips.

Quinn interjected quietly: "Earth, being the smaller planet and already set up by other problems such as the *greenhouse effect* will tilt first—the global shift in effect—trigged by Jupiter . . . is that the correct surmise?

LaMont's head nodded in silent agreement.

A sharp knock at the bedroom door and the sound of the knob turning caused Quinn to rise from LaMont's bedside.

Karin slipped into the room, face hard with anger. "Hartlett's on the screen. Kubedera's with him with a gun to her head again."

Quinn turned to LaMont. "Can you handle this?"

"He'll kill if I don't give him the answer he wants." LaMont attempted to rise.

Reaching down, Quinn picked up the geophysicist in his arms. *Can't weigh more than ninety pounds.* With Karin's help, he placed LaMont in the wheelchair, tucked the robe in and pushed him into the

143

computer room, aligning him with the screen. He turned to his men who sat on the floor, backs to the wall and held up his hand for silence.

Karin flipped the audio control button on. "Doctor LaMont is here but he should be resting. Please don't do anything to upset him."

Quinn was impressed at Karin's gutsy, straightforward approach.

"The good doctor can rest for eternity Nurse Maki—if he confirms what I suspect—that our little Japanese wonder here has introduced a *gremlin* into the program."

Hacking, LaMont cleared his throat and spoke into the transmitter.

"Take the gun from Maria's head. Let her talk to me."

Hartlett pulled the gun down and positioned the software engineer directly in the eye of the camera.

"Maria, Bochert's dead—you don't have a team leader. I'm in charge again. Remove any artificial barriers immediately. Let the run proceed. Do you understand what I'm saying?"

The young woman nodded, tears flowing.

"Good, Maria. Get back to the computer room and clean up the program."

A hand reached out yanking the woman from the camera range. A moment later, Hartlett reappeared. "A very wise move, Doctor LaMont. There is nothing to be gained by creating further obstacles. In fact, I am taking steps to make sure you or someone else doesn't interfere with our efforts here in the mine or on the surface. It appears that conflicting orders have been given to the tank crews. If that is the case and my men come to harm, *two* of your people will die for each *one* of mine. Pay close attention Doctor LaMont, Nurse Maki and anyone else who may be watching. Paired up, your people will appear before the camera in the exact order they will die should it come to that."

Hartlett flashed a cynical smile and touched his field cap in a mocking salute. Nick, Wally, Lubo and Running Fox came to their feet. Concerned that one of them might speak, Quinn quickly turned off the two-way speaker. Two-by-two, men and women held hostages were paraded before the camera. As the tenth pair appeared, Nick cursed. "Stinking, rotten bastard . . . .!"

Mrs. Pavo, the pudgy day-shift cook was bound up with an attractive, dark skinned, Native American. Running Fox said nothing as his sister and the Finnish cook were pushed in front of the camera,

144

made to stand for several moments, then ordered away by a voice unheard. Torment and despair was mirrored on each face paraded before the camera. When the last couple made their appearance, the screen went blank. Nick, face grim, silent, shuffled back to the wall and sat down, Wally beside him, a comforting arm around his grandfather's shoulder. Lubo, grimacing in pain, took his place on the floor, head shaking.

Running Fox, war stripes still on his face, remained standing, dark eyes fixed on Quinn's. When he spoke, his voice was without recrimination. *"Ogan* still lives."

Quinn stared at Running Fox and then the other volunteers. Nick's head bobbed. He suddenly looked his age and Wally, was just a kid again. Lubo, face drawn, had his eyes closed in pain. He looked at Karin and saw dark shadows under her eyes for the first time. She too, appeared to be strung out too far. Something was needed to break the impasse. He turned to Running Fox. "Agreed but I call the plays."

The former SEAL headed for the basement.

Quinn took Karin by the arm, leading her into the kitchen.

"I'm gonna' try a long shot Karin, Running Fox is right. Your presence *might* speed things up. If it gets too bad you can leave. You willing to be part of this?"

Eyes glazed with exhaustion, Karin nodded and moments later was guided down the back staircase. In the dark root cellar, the young Russian, shirt off, hands tied behind him, sat on a potato sack. Quinn had Karin sit on the wooden stairs where her eyes would be level with those of the Russian.

"Get him to his feet Jimmie!"

Running Fox yanked the prisoner up.

"Show our young friend *'Ogan'* again."

Reaching into the pail, Running Fox held up the snake pickerel in both hands. The mottled green and black foot-long fish twisted violently in the Indian's grip.

"Make sure he gets a good look!"

Running Fox put the fish's head inches from the young Russian's face. Eyes growing larger, the prisoner stumbled, falling back against the trunk that earlier had supported his body.

"What's he going to do with the fish Caleb? What's going on?"

145

Running Fox jerked the Russian upright again and pushed the snout of the sharp-toothed pike under the man's chin. Quinn purposely ignored her question.

"Have him stand in front of the trunk, Jimmie–cut the thumb ties." He spoke to Karin, but his eyes never left those of the prisoner.

"Our friend knows what Jimmie will do with the fish Karin. If he answers my questions, Running Fox does nothing. We know he speaks English and that's what's required here."

Feet apart, hands on hips, Quinn looked the Russian in the eyes and then barked; "DROP YOUR PANTS SOLDIER!"

The young soldier's hands went to his mid-section but he hesitated, gaze fixed on Karin. It was the moment Quinn was betting on. He moved toward the prisoner, making sure Karin remained in the prisoner's field of vision. His voice was softer now, so low Karin could not hear the muted conversation. "You understand *me* and you understand what will happen if you don't answer questions. You'll drop your pants, bend-over and re-assume the position over the trunk. Running Fox will proceed to insert the fish's head in your rectum. What happens then, Jimmie?"

Running Fox put the fish on the Russian's shoulder, letting its ice-cold tail slap the man's neck. The jaws snapped open, rows of small, razor-sharp teeth showing. The man's head bent away from the fresh water pike.

Taking a clue from Quinn, he too spoke softly, "Once its head penetrates beyond the sphincter muscles—in past the gill plates—it can't be taken out. There's only one way to go—into the intestines, snapping at whatever blocks its progress."

All three men heard Karin suck her in breath audibly. Ears straining, she'd picked up their hushed conversation.

"Most interesting Jimmie. And you, Russian what are your thoughts?"

Quinn gripped the man's chin, forcing his gaze away from Karin, making him look directly into his eyes. He moved within inches of the man's face, one hand coming to rest on those of the prisoner who still gripped his belt buckle, voice a whisper. "You have a choice soldier. Drop your pants and let the young lady watch the ceremony or—answer a few simple questions."

146

At that moment, Jimmie placed the cold, slimy fish against the man's cheek, holding it tight against the skin.

Eyes rolling wildly, the young prisoner blurted, "I speak! I speak!"

Quinn pushed him onto the trunk. He motioned Jimmie to put the fish back in the pail but moved the bucket with his foot to keep it sight of the Russian.

"Name?"

"Sergeant Specialist Vadim Solomenstev."

"Your primary assignment?"

"Explosives."

"You served in Afghanistan?"

The prisoner blinked. "Yes."

"How long?"

The man hesitated

"How long?" Quinn asked again.

"Two . . . two years."

"You don't seem very sure Sergeant Specialist Solomenstev?"

"I..I am sure . . . two years."

"How many SPETSNAZ are involved in this operation?"

"Twenty-eight I know of. There could be others—many others."

Quinn shook his head grimly at Running Fox.

"What unit did you serve in?"

Solomenstev looked blank.

"Company, brigade, division—what specific group were you in?"

The sergeant specialist stared at Quinn until the pike once again came to rest on his neck.

"I am *reydoviki*—raider in your words. My unit is called *Ikhotniki* and we are known as the '*huntsmen*.'"

The slight tilt to the young man's head told Quinn he was proud of his unit. *Got to remember that one. Raiders and huntsmen.*

"Your commander's Russian name?"

The prisoner swallowed hard but said nothing.

"No unanswered questions, soldier."

Jimmie reached into the pail and rubbed cold water on the man's neck in soft, almost sensuous manner.

"Colonel Leonid Arbatov."

"Did you serve under his command in Afghanistan?"

147

The prisoner blinked his eyes and when he opened them, he stared above Quinn's head. Pattern established. Quinn knew he had the man. He motioned to Jimmie and they retreated to the stairs where Karin sat transfixed by the interrogation.

Whispering, he spoke to Karin and Running Fox.

"We're getting answers but I don't think I know the right questions. There's something he's covering up about Afghanistan and damned if I can figure it out—any ideas?"

"Are you going ask him about his back?" asked Karin.

"Good idea," nodded Running Fox. He's been whipped or he's whipped himself."

Quinn rubbed his brow. It had never occurred to him that the wound might be self-inflicted. *Running Fox is like Mumford. Looks at situations from different perspectives.*

"That's it then—next on the list. Maybe it'll lead someplace."

Suddenly the house shook, dust fell from the wooden beams onto the dirt floor. Three more blasts followed in quick succession.

"Damn it to hell! The tanks have opened fire!"

He turned to the prisoner. The fear was gone from the man's eyes, replaced by a thin, hard smile. Quinn knew he'd lost the upper hand, psychological momentum untracked by the distant cannonade.

"Your back, Sergeant Specialist. It's covered with scars. How'd that happen?" The man's shoulders seemed to straighten. Another volley of cannon fire thundered overhead. Dust filled the air and the dim bulb swung from the beam. Quinn knew further questioning was useless. Discouraged, he turned to the Ojibwa.

"Tie the son-of-bitch up Jimmie. Loop a rope from his thumbs to his feet and then hoist his legs off the ground. Turn off the light and let Solomenstev think things over in the dark. I'm heading up and see if I can get Holmes on the radio." He eased Karin ahead of him, the two of them racing upstairs.

In the computer room, LaMont, ashen, was at the console. He turned to Quinn the moment he came through the door.

"It's General Henderson, says it's urgent. I told him there was gunfire coming from the mine!"

Quinn took the headset as Karin pulled LaMont's wheelchair back. The house shook as another volley was fired.

148

"GATEKEEPER here! All hell's breaking loose again, Jocko. Make this quick!"

"Hell's about to happen back here, Caleb. Russian's seem poised to go into a full dress, war-on-all-fronts, alert."

Quinn closed his eyes. "Regarding the prisoner, here's some, rank and serial number." He rattled off the data including the reference to "*huntsmen,*" "*raiders,*" and the Colonel's name. "We ran out of interrogation time. There seems to be something about Afghanistan he's trying to protect. Hartlett must know something's happening topside. Tanks are firing and I need to contact my man inside the Guard."

"You made contact with the Guard Commander?"

"No but we tracked down my reconnaissance leader and he's in charge of trying to pull the Guard away from the Russians. Time, General. LaMont estimates the computer runs will be ready as early as eight AM. Hartlett's promised to kill two hostages for every man he loses. The killing might have already started."

Cannon fire continued in the background. A window shattered on the north side of the frame home. Running Fox left to investigate. Karin helped LaMont out of the wheelchair and onto the floor.

"Somebody better be thinking about the townspeople, General. Innocent people could get killed real quick up here!"

"Somebody is Caleb. The Highway Patrol, local police and sheriff's department are starting a house-to-house clearance. They'll try to get everybody out of Soudan within the next two hours."

"One last request, General. Can somebody track down aircraft and pilots used here in Northern Minnesota to fight forest fires? I want access to fire bombers or helicopters rigged to dump beau coup water damned quick! I'd start with the Department of Natural Resources in Brainerd and go from there."

"Unusual request, but consider it in the mill, Caleb."

Running Fox dropped a metal fragment on the desk. Quinn picked it up gingerly and shook his head. *5000 of these razor-sharp little gems per round—fired at innocent soldiers on peacetime maneuvers—in their own country.*

"APERS being employed, Jocko—GATEKEEPER'S signing off."

Searching for the scrap containing the radio frequencies, he glanced around the room. The gunfire had acted like a shot of

149

adrenaline. Nick's eyes sparkled as he fitted a new clip into his weapon. Wally had his rifle laid out ready to go. Lubo, arm in a sling fashioned by Karin, was cradling his weapon. Running Fox, in war paint, stood with his back to the wall, hand gripped around his lever-action Winchester.

Slip in hand, Quinn set the radio frequencies to the first channel given by Holmes. A steady hum greeted his ears. He switched to the second and then the third. Same results. There was no response. He clicked to the standby channel. "Anybody got ears on out there? This is GATEKEEPER Comeback."

The radio crackled. "We got ears. Who's out there? Comeback."

Quinn grinned and repeated Khe Sahn password; "Last One Out—turn off the lights."

"Colonel, this is Captain Holmes, bloody but unbowed."

"What's happened, Ernie? Fill me in."

"I patched orders out to the men on the line—maneuver was being cancelled because of the unexpected weather and risk of death due to carbon monoxide poisoning—all Guard personnel to make their way down from the mine to the high school. I figured that would get most of them out of range, but don't know how well it worked. The Command track issued an order countermanding mine. Minutes later the tanks unloaded. It's confusing up here but the shelling seems to be over. We had to leave the Dry House when the Russians opened fire from the Head Frame and Engine Room. We got the RAT rig out and set up— freezing our asses off in the . . ."

"Hold the location, Ernie. Don't call down fire on yourself. Two questions: how long can these tanks run at idle? And where is the fuel supply?"

There was a pause. "Twelve hours at idle and they're six hours into their supply. APC with barrels and pump is scheduled to make the run to each tank at daybreak. In this weather, the tanks need to run a constant 1200-RPM. If they use searchlights, they'll be forced to run even higher to keep batteries charged. The fuel depot is in the center of the perimeter, which makes it south and slightly west of the Crusher. I'd guess about one hundred yards away."

"Can you see anything from where you're at now?"

"Negative. When the snow stops, we should have a good view of the area."

"They must be up in the rocks northwest of the mine."

Quinn looked at Running Fox and nodded.

"Ernie, can you hole up there and survive this cold?"

"If we dig in we can make it."

"Good!" He scanned the frequencies available. "Guard a new channel, Ernie. Remember the dimensions for a regulation slit trench?"

* * *

The President held firm to initiating the Hot Line, adamant that the armed forces stay below the "Nuclear Alert" level.

Within minutes, the sleek 747, distinguishable from other "heavies" by the second "hump" on its topside, was towed out of the cavernous hanger at Wright Air Force Base in Ohio and in pre-dawn darkness, became airborne with priority vectors to Andrews Air Force Base in Washington.

Resting on a cot in the Situation Room, the President experienced a fitful, half-sleep.

Secretary of State Weller, CIA Chief Yoder and National Security Adviser Hart worked on a draft of the Hot Line text being sent to the Russian President and nominal leader of the Commonwealth. In a windowless room in the heart of the Pentagon, a sergeant and a colonel, both fluent in Russian, readied the even-hour message to be sent over the Moscow Link.

Precisely three minutes to the hour, a civilian typed in the prescribed combination of keystrokes, indicating Moscow was ready to receive its hourly test message. On the hour, the American sergeant sent the text indicating a second message would be sent within thirty minutes and was for "*Eyes Only—President.*" The transmission indicated photographs would be sent as well.

The President was nudged awake and given the text hammered out by his staff. Brushing sleep from his eyes, sipping coffee, he made several pencil changes and authorized its release then made his way toward the end of the Situation Room where Generals Henderson and Smallzreid studied the latest reconnaissance photos. The marine general handed a black and white blow-up to his Commander-in-Chief.

151

"Colonel Quinn didn't want ordnance dropped so we sent photo RECON in to help with the sound effect. The circles are tanks, running at idle and giving off heavy heat signatures." He pointed to a similar mark in the middle of the ring.

"Fuel trucks. He wants a surgical laser-strike take-out of the dump within the next forty minutes, before the tanks can be re-fueled. Benedetto's people are preparing a strike. Quinn's requested we locate and make available any aircraft designed for fire-fighting work. We've got three. A converted DC-6, a modified B-24 and a Bell Jet Ranger rigged for water drops. Pilots are available and standing by. He hasn't indicated what he has in mind Mr. President, but said the idea was given to him by somebody called Nick."

The President studied the grainy infrared photos. The juxtaposition of state-of-the-art laser targeting and use of a fifty-year-old bomber brought a glimmer of a smile to his face.

"How is our man doing up there?"

Tapping the dark outline of the Dry Room, the Commandant looked up. "He and a volunteer located the surface COMMCENTER, took out two Russians and gained some radio gear—found a former enlisted man who served under him in Vietnam—now a Guard Captain. Captain and two enlisted men pulled the radio out under fire and have relocated somewhere on the perimeter. From there, the captain gave orders to cancel the maneuver. Minutes later, a firefight broke out. That's where it stands."

"A fire fight," mused the President. "Any casualties?"

Henderson pulled another photo from the stack placing it on top of the first.

"Low-pass infra-red, Mr. President. Thirty minutes old."

There were a number of small dots circled with red. Eyebrows arched, the President tapped one of the dots with the gold pen.

"And?"

"They're Guards who followed the first order and left the tanks. This photo was taken as they made their way down from the mine to the local high school. If they don't show up on the next few photos we'll have to assume they've frozen to death—or, if wounded, they made their way out of firing range."

The President's hands slapped palms down on the photo, head dropping between his shoulders.

"Probably more dead below ground, Mr. President. The Russian Commander ordered two hostages killed for every one of his men taken out. At LaMont's end, we have an interesting problem. The man is in the terminal stages of AIDS and out of medicine—and what medicine he has might be placebos. Seems he's part of joint study with University and the Disease Control Center in Atlanta."

The President's head snapped up and he rose to his full six foot, five-inch height. "How long until the 'Eyes Only' text goes out?"

Henderson glanced at the digital clocks.

"Seven minutes, sir."

"You think the our man will be there when the message comes in?" The President's gaze moved from General Smallzreid, back to Henderson. Both men nodded.

"Good! Give me that damn text. I want to make some last minute changes!"

General Smallzreid turned, heading for the transmission console.

The President gripped Henderson's forearm.

"And regarding help for Doctor LaMont, you, General Henderson, get on the line to Atlanta and tell them to pony up what's real or somebody will be researching canker sores in Killer Whales . . ."

*   *   *

Quinn gave Karin last minute instructions on using the radio and turned to leave. He'd put the doctor, alternating between profuse sweating and uncontrollable shivering, into bed, last of the IV dripping into his shrunken veins.

"When will I see you again?"

Karin grasped Quinn by the w.sts as Lubo, last man down the stairs, turned and lifted his rifle in a salute. In an instant, his bulky form was swallowed up in the darkness.

"The next three hours Karin—we either stop the computers or the transmission. To accomplish the first I've got to get down into the mine. The Russians own the tanks, access to the Head Frame and Engine Room—and the shaft going in."

Karin pulled the white hood over Quinn's head.

"Keep your head down, Caleb. No dumb Finn stuff!"

Quinn brought her into an embrace, his lips brushing hers. She rubbed his grizzled chin with her hands and then drew away and disappeared down the stairwell.

At the Land Rover, he divided the dynamite evenly between the men, each volunteer taking twelve sticks along with blasting caps and wire. Nick took the point as the small band moved out, white apparitions mushing through waist-high snow. The earlier path to the Alaska was already blown over.

Just beyond the last house, now vacant, Nick suddenly dropped to his knees, head barely above the snow line.

Quinn and Running Fox moved up beside him.

"Jahah!" whispered the old man, his breath coming in short, jerky puffs. "We gots company."

Staring into the blowing snow, trying to establish a reference point, Quinn shook his head. "Can't see shit, Nick. Must be snow ghosts."

Running Fox tapped Quinn's shoulder.

"Eleven o'clock, near the tallest pine. I'll circle around."

Patting the Ojibwa on the back, Running Fox faded away. Quinn glanced at his watch. If Henderson delivered the next flight on schedule, he wanted to be in position when the fuel trucks came under fire. Nick, Wally and Lubo hunkered down in the snow next to him.

Minutes later, Running Fox appeared directly between them and the large pine. Behind him, two men in white, supporting a third, plowed through the snow.

"Guard, Colonel. They've got a wounded man."

Quinn stood up.

"Get'em up here! Nick, do the drill. Check'em out quick."

The two men lowered the third in the snow between Quinn's men.

Wally pulled the wounded man into a prone position as Nick, speaking rapidly, fired questions at the panting soldiers.

Probing with a penlight, Wally looked up.

"Shot through the back, sucking chest wound, Colonel. This guy needs help. Bad."

Nick turned to Quinn.

"Guard, Colonel, and scared shitless!"

"Scared, man, but not shitless!"

The taller of the two Guardsmen edged toward Quinn.

154

"I'm Sergeant Maki. If you guys are going up there, I want in. Two of my crew got blown away by their own damn tank!"

Quinn, clasping the man's shoulder, felt a heavy belt under the white parka.

"Got a radio with you?"

"Yeah, a PRC 7-7"

"Well, Sergeant Maki, you're gonna get your wish. I'm Colonel Quinn, USMC and I'm running the shooting gallery for our side. Those are Russians who killed your friends and we are definitely going back up!"

He turned to the logger.

"Lubo!"

"Yahvold!"

"You and the Guardsman take the wounded man back to Karin. She can call for help from the police. If you're up to it, come back following the trail. You'll end up under the lip of the Alaska shaft."

He turned to the man kneeling over his fellow Guardsman.

"We're short of firepower. Give your weapon to Wally and then get your friend on his feet. If he wants to live, he'll have to walk out of the woods. Now move it!"

The young private scrambled to his feet, pulling the gun off his shoulder, yanking clips from his parka.

They got the man up. Lubo took most of the weight as they headed back to the frame command post.

"Jimmie, up front! I want Nick back here while we move up the ravine. Keep your eyes open for other strays."

Running Fox turned and moved toward the dim outline of the cut ahead, his white sheet snapping in the wind.

Sticking close to the sergeant, Quinn peppered the man with questions about the defense perimeter, condition of the tanks and most important—where Russians on the surface were located. Moving as quickly as the heavy snow permitted, they slipped into the rock depression that formed the opening to the Alaska Shaft.

In the lee of the howling wind, the men huddled around Quinn.

"Time and terrain doesn't allow much in the way of variables. If the Air Force delivers on time, we'll have fireworks at seven-fifteen. I want all of you to be in firing position before they deliver the coup de grace to the fuel depot. Nick, you stay close to me. You're my guide to the nether world. Sergeant Maki!"

"Yo!" The muscular Finn pushed closer.

"You get to go back where you came from. You know the approximate location of the tank with the ground radar. See if you can take that out. "Jimmie, go back around the long way like you did on the first pass. On the ridgeline you should find Captain Holmes. Take his two men and spread out at fifty-yard intervals. When the dump goes, that's the signal to put sporadic fire into the Engine Building. The objectives: Keep the Russians looking over their shoulders at all times with their searchlights on. We keep those Russians at the Engine House occupied but not able to run the cage mechanism. Wherever possible, we cut their communications lines. At this point, everything we're doing is for one purpose—to get access to the cages so that one or more of us can get into the mine. Nick and I will try to neutralize the tank at the Head Frame, then move on to the Cable House."

He looked around at the still forms, faces recessed deep in white hoods, with ice on beards, eyebrows and mustaches.

"If the dump goes on schedule that's when everyone moves." He reached for the field radio and selected Channel 16—measurement of the standard slit trench—one foot by six feet.

"KHE SAHN, GATEKEEPER here. Fire storm coming down at 0715 hours. KHE SAHN you should have company behind you in fifteen minutes. He's friendly and will take your two men. Maintain your position and keep an eye out for the missing APC—visually or by radio chatter."

"KHE SAHN confirms."

Quinn rose, facing his small band.

"If we can isolate the Russians from their topside commander and the big cheese below, they might be easier to deal with. Last instructions; Dry House is first priority rendezvous, this hole in the ground, second. If you can't make it to the first come here. Move out and stay low to the ground."

Running Fox was first out of the rock hole followed by Sergeant Maki and Wally. Quinn grabbed for the bags containing grenades, bow and radar unit. Quickly, he lashed the compact radar unit to the twelve-volt lamp, wiring the electronic device to the batteries. Behind him, the

156

old Finn wired dynamite sticks together in bundles of three. Sliding down into the snow, Quinn pulled the sheet around him. For one brief instant, he thought he heard the sound of fire trucks somewhere down the hill.

*   *   *

Tapping his fingers on the green felt surface, the Russian President noted the stack of reports marked, "SOVERESHENNO SEKRENTO" had increased from a dozen to over fifty. More were delivered every few minutes by a KGB general. Field Marshall Korovkin sat stiffly. Manarov had his tailored suit coat off.

KGB technicians had completed the relay between State Headquarters and the command post set up outside the Green Room. Below the Kremlin, at Sverdlov Square Station, the Blue Train had been ordered into position. A light snow fell in Moscow Square. Behind the seamless steel wall, the special train was already being boarded. Military and civilians disgorged from elevators linked directly to the Kremlin above. Color-coded *propusks* allowed them entrance to a car having the same thin band of color beneath the windows.

No one was allowed anything more than a briefcase. Priority passengers in the last car in line, color coded blue, would be shunted off ten miles from Central Moscow to the surface at Khodinka Central Airfield for flights to underground posts, others to Mayakovskaya or Byelorusskaya airports. To prevent undue commotion and rumors, strict silence was enforced by GRU security officers.

In the Green Room, the first of two facsimile machines hummed into operation. Seconds later, the photo unit blinked green signaling the start of transmission. Manarov, startled, dropped his cigarette onto the green surface. He started to rise but the President motioned him to remain seated.

"Put yourself at ease, Viktor. I will be the first to read the text." Moving quietly across the thick carpet, the President stood before the unit stroking out the plain copy text in perfect Russian. He smiled at the warm greeting from a man he'd come to admire. They had much in common, thought the Russian leader. As the text rolled up, the Russian's face lost its smile, his hands gripped the edges of the softly vibrating machine. Thankful his back was turned to his ranking military leaders, he glanced at the second machine and saw the photographs

157

referred to in the text materialize. Slashes indicated the end of the text but more data appeared in the form of a short listing and then, character generation ended. The second machine purred for several more seconds and it too, came to a halt. Reading the message twice, he tore the paper from the roll, studying the photos. Experience, going back to childhood, told him the birthmark on his forehead was several shades darker than normal. He willed himself into a condition of mental constraint and then, returned. He paused directly behind the KGB chief who realized from the intense look in the Field Marshall's heavy-lidded eyes that something unusual was taking place. Unable to stand the tension, KGB Chief Manarov swiveled around in the leather chair to face the head of the Russian government.

Holding the two documents together, the Commonwealth leader said nothing and continued to his chair at the head of the table. A sharp rap at the door broke the silence. A major general, bearing more documents entered the room. Turning to face him, the Russian, a steely edge to his voice, announced, "There will be no further interruptions."

Startled, the stocky general in the brown uniform with heavy red piping looked to the Field Marshall for confirmation.

The President saw the furtive glance. "Leave immediately! When your presence is required it will be requested—by me!"

Face flushed, spinning on his heels, the general closed the door quietly.

Laying the papers face down, hands clasped, the President looked at Manarov, left to the Field Marshall and then, slowly removed his black-rimmed glasses, folded them with deliberate slowness, tapping his compressed lips for several long seconds. "*If*," the Russian leader let the word hang in the heavy silence for a moment, "*I were President of the United States*, I, too, would have ordered out my military apparat—given the circumstance outlined in these messages." Leaning forward, both elbows on the green baize surface, he continued, "I am surprised Field Marshall Korovkin, General Manarov, we haven't seen even greater preparedness for war." Standing up, he took the text portion of transmission and slammed them down in front of the supreme military commander of all Russian forces.

"You, Korovkin, control the SPETSNAZ, a force I am beginning to find a millstone around my neck!"

He half-turned to the KGB chief and showed him the photographs of the two captured Russians in Washington, both dead. Pictures of the disks, whole and apart, accompanied the photos of the man and woman. The list contained the names of the troops already identified in Washington and two known to be in Northern Minnesota—Colonel Leonid Arbatov and Sergeant Specialist Solomenstev of the *voiska spetsialnovo Nasnacheniya.*

The President sat down in the high-backed chair. "Please be so kind as to read the document and review the pictures."

Manarov and Korovkin, exchanged documents in silence.

The Field Marshall, forehead tinged red, neck swelling within the stiff collar, stared at his KGB counterpart, eyes growing porcine as the thick lids compressed.

"The SPETSNAZ are under your command?" asked the President.

"Da!" the Marshall's guttural response.

The chair swiveled. He faced the KGB Chief. "And you, Viktor, you have access to them, can deploy them for the good of the State?"

Manarov nodded his answer.

Putting his hands palm down on the table, the President rose, about to speak. Conscious of his throbbing forehead, he straightened up, abruptly turning away, walking toward the window.

Forcing himself to concentrate, he a subject matter close to his heart, Peter the Great, his personal hero. He mentally reviewed the man's accomplishments in chronological order. The ploy worked. The pulsing within his skull lessened. He knew either man sitting behind him could wrest his leadership away with a snap of a finger, citing stress or weakness in the face of enemy threat. The President was also aware of the peculiarity of the Commonwealth's Constitution—still in the formative stages but carrying over the clause that would put his country under the iron fist of the *STAVKA*—in peacetime, a paper hierarchy composed solely of military leaders, Field Marshall Korovkin at the top. The moment the decision was made to engage in war, Korovkin would assume total and complete control of the government. At the pinnacle of the *STAVKA*, Korovkin would have, in his hands alone, concentrated powers last held by Josef Stalin.

An involuntary shudder raced down the President's spine. *Russia will return to the Stone Ages! No. It cannot, I will not allow it to happen.* He lowered his gaze.

Slowly, in full control, now, he faced the two men who, with him, held the future of Russia in their hands.

"The President of the United States has asked—*no*—demanded that I personally respond to this flagrant violation of the sovereignty of his country. Were the conditions reversed, I would insist upon the same."

He looked long and hard at the two men and then, choosing his words carefully, spoke. "Field Marshall Korovkin—General Manarov, who is responsible for sending the SPETSNAZ into the USA, wearing American uniforms, penetrating an underground research facility, killing innocent civilians and soldiers in the process?"

Neither man spoke.

Knuckles curled under on the baize surface, he leaned forward and spoke in a virtual whisper; "He has asked for my response before the hour is up."

He straightened up.

"AND BY GOD HE WILL HAVE IT!" His voice thundered and his balled fist slammed down, bottles of mineral water fell like bowling pins. Dried pastries flew from the plates.

Korovkin was the first to respond. He did not look at the Russian leader, burning gaze instead, riveted on his military counterpart, KGB General, Viktor Manarov.

"The Army, Mr. President, has not authorized *any* such operation. There are over 36,000 SPETSNAZ under my command. Not a single one has been cleared for such a mission. *If*, a mission such as you describe has actually been arranged, execution underway, it has been done without my knowledge and—in complete violation of protocols existing between the military and civil government. This I will swear to in front of all twelve members of the presidium who wait in the chambers below." The Field Marshall looked at the Russian Leader without blinking and continued; "If—some other branch of this government felt it necessary to initiate such a mission, it did so without my authorization or knowledge. The Russian Military will not be the goat for such baseless accusations!"

Hands tightly clasped, thick thumbs tapping together, Korovkin resumed the intense scrutiny of his fellow officer.

The President turned in his chair.

"General Manarov, it appears we need to hear from the *Komitet Gosudarstvennoi Bezopasnosti.*" His voice was quiet.

Manarov shot a furtive glance at the Field Marshall. "It is a given—the KGB has, in the past, used the SPETSNAZ in specific operations. Dozens have taken place in Afghanistan, Angola and Central America. Individuals have been assigned to the U.S. for social and geopolitical indoctrination. The Americans know some members of our Olympic teams have dual roles. We do not admit that but it is a fact hard to hide from close scrutiny."

Reaching for his cigarettes, Manarov had second thoughts, tapping instead, yellowed fingers on a tar-rimmed ashtray. Appearing to take a shallow breath, his shoulders arching slightly, he faced the leader of the disparate Republics. "The KGB has *not* authorized any mission such as the Americans have described. Your intelligence apparat still has hundreds of thousands of operatives at its beck and call worldwide. I don't claim to know where each of them is at any given moment, but I can assure you, there is no KGB activity going on in the United States having any connection with my agency and Korovkin's SPETSNAZ!"

The President pushed back in his chair, his gaze shifting from Manarov to Korovkin and then, in exasperation, toward the portraits of his country's fabled founders and leaders. Nothing in their painted eyes offered the Russian Leader solace or solutions. The small conference room was tomb-silent until Field Marshall Korovkin reached for the President's message, carefully tearing the listing portion of the main body of the text. He took the facsimile photos in front of the KGB Chief and stood up.

"Perhaps someone is lying—perhaps these men are agent provocateurs. We shall see. It would not be the first time in history the enemy has created an incident in order to justify its actions."

Seconds later he was at the door, conferring with the four-star general serving as a courier, moving incoming intelligence into the room, messages to the seated members of the presidium and the nine alternates, one floor below. The flag officer nodded, taking the papers from the Field Marshall.

Returning to his chair, Korovkin tilted back, huge peasant hands drumming the table. "We gave them glasnost—and they give us this. My earlier recommendations still stand. I say we extend our military flanks—buy time and space in which to operate—poll our commonwealth members to see who will stand with us."

The President stared at the Commander of Federal Forces in stony silence then turned to Manarov. The KGB chief's bobbing head indicating concurrence with the Field Marshall. Voice resigned, the President asked of Korovkin, "How long before you have something on the SPETSNAZ?"

"Ten minutes, Comrade, er, Mr. President. Ten minutes."

The Commonwealth leader looked up at the three clocks giving Moscow, Washington and Zulu Time. *Forty-seven years a survivor. And now everything I am, everything I desire to be, everything I seek for my country could end in a matter of hours ....*

*   *   *

Mossad decoders in Tangier, with the help of an IBM computer, had the English version of the text twenty-six minutes after tap—a re-coded Hebrew version relayed to General Eli Ben-Yaacov, on inspection tour at Dimona, Israel's nuclear weapons facility in the Negev Desert minutes later. Within minutes the wiry head of LEKEM and affectionately called *baleboss*—Yiddish for "*the superintendent*", chose to return to Tel Aviv and his subterranean headquarters on the edge of the Ben-Gurion airport. He placed one call over the secure line at Dimona advising the Premier to join him. Already under siege from militant splinter parties on the far right and left, the Premier at first demurred but acquiesced when Ben-Yaacov repeated a phrase recently cited on the floor of the Knesset, "When bull and bear go head-to-head, lesser animals are sure to wallow in the gore."

Ben-Yaacov left Dimona pleased with the inspection. Israel's underground nuclear facility had increased its inventory of 20-kiloton Jericho weapons to 72 and had perfected their use for air-to-ground and ship-to-shore delivery. Patriot missile launchers left from the 100-Hour War ringed Dimona as well as other critical sites. Minutes after suiting up, Ben-Yaacov taxied onto the four-lane section of highway doubling as Dimona's airstrip, eased the throttle forward and watched the ground speed indicator accelerate rapidly. Ahead, in shimmering heat, sand and

162

sky seemed as one. At 120 knots indicated, rotation speed, he pulled back on the stick, bringing the powder blue, single-engine *Kfir*—dubbed Lion Cub—off the deck and hit a switch bringing the wheels up with an authoritative, "thunk!" In minutes, he was between the low mountains, heading northeast at 480 knots.

*   *   *

Waiting for the Air Force to make its contribution, Quinn helped Nick tape dynamite together in bunches of three. Snow eddied into the spots vacated by his departed troops. The work required bare hands and both men had to stop frequently, warming numbed fingers. Snow gusts swirled between the buildings accompanied by shrieking winds.

Suddenly, the sound of crunching snow could be heard. A huge white form huffing shuffled up the narrow trail from below. Quinn reached for his weapon.

"Lubo!" a voice hissed out of the dark. Then the giant lumbered into the snow-packed well. Lubokowski's bulk was followed by another, smaller figure. Straining to see in the blowing snow, Quinn realized the soldier was wearing a regulation white parka. As the man came forward, he saw the dark hole in the chest behind the shotgun.

Lubokowski laughed. "I tried, Colonel—said it was dangerous."

Quinn pulled the soldier down, yanking the hood back. He stared into Karin's face. "What in God's name are you doing up here? Who the hell said you could leave LaMont?"

"Jahah!" sputtered Nick, making room for the new addition.

Karin pushed the hair over her forehead. Quinn's nose was inches from her face, anger in his eyes.

"LaMont died holding my hand. A few minutes later, I heard sounds down below. When I opened the door leading to the kitchen, I could see smoke coming up from the cellar. I thought the prisoner might have gotten loose, started the fire and escaped but I heard him scream down there. It was . . . ."

Quinn could feel the young woman tremble.

She looked up, tears filling her eyes.

"I ran back to LaMont's room and for a minute just stood there. A second later, flames where already coming up the back stairs. I couldn't carry LaMont's body so I . . . ."

163

"Don't worry Karin, maybe cremation was in the cards all along for both of them. You couldn't have done anything more."

"I called Chief Heitala and he told me to get out. When I got outside Lubo, the wounded guardsman and his friend were coming out from the trees. We waited until the fire truck and police came. Then Lubo said he had to go back, that you needed all the help you could get. I took the wounded man's parka and grabbed this shotgun from your Land Rover—I need to help."

Quinn shook his head from side-to-side.

"Just after LaMont died and before the fire stared, Hartlett came on the video. I told him what happened. He said it didn't matter, that the computers were running. He wanted to know where his man was. I said I didn't know. Next thing I knew, the camera focused in on six bodies lying on the ground—laid out side-by-side."

She grabbed his wrist. "He's killing them, Caleb, killing innocent people!" Her voice rose.

"Shh! Karin, we've got Russians just over the rocks."

Karin bit her tongue, regained her composure. "I couldn't stay Caleb. I'd say the wrong thing and he'd kill more because of me. There wasn't anything to do for LaMont or the young soldier. I . . . thought I could help here."

"Jahah! She's a *Lotta* for sure Colonel!" The old Finn chortled.

"What in hell's a *Lotta*, Nick?" asked Lubokowski.

"She be a donut girl, nurse and saint wrapped up in one. Gots her name from a young Finnish sweetheart who followed her man into battle way back when. Yes sir! We got us a genuine *Lotta*!"

"And I've got good news!"

Quinn stared into Karin's hooded face, wondering what he should do with her, silently admiring her spirit, angry she was on the edge of a fire fight about to take place.

"I'll take the good news," he said gruffly, trying to mask his feelings.

"Your General Henderson said they've figured out a communications relay system. Something called a "WAC" is coming on station in twenty minutes. As long as it's needed, one will be airborne overhead."

"A WAC?" Quinn thought to himself and then laughed.

164

"It's *AWAC* Karin, and that's damned good news! How do we communicate?"

"Make contact with Holmes, they'll guard all channels. You and the Captain can pick out which ones you want to talk on."

"Anything else?"

"Yes! They'll move anytime you want on the satellites. They've got only two missiles in position however. You'll have to advise if they should be taken out if you can't stop Hartlett on the ground."

Quinn digested the information in silence. He didn't have an immediate answer.

"Also, Henderson said to tell you, '*we've lit the wick under the Hot Line*,' whatever that means."

He looked around in the dark, faces so far back under white hoods only puffs of vapor appeared. "It means Karin, Nick, Lubo, that the President's taken the initiative in talking to the Russians. It's one way of preventing the outbreak of full scale war."

He glanced at his watch.

Karin's hand grasped his forearm. "There's a couple of more things I've got to tell you."

Rolling on his side, he pushed snow away from the rock and drew Karin closer.

"We've only got a few minutes. What else happened back there?"

"The Russian, Hartlett, could he be a priest of some sort?"

"Huh? He pulled her even closer. In the snow and darkness, her eyes were black pools.

"When the camera showed the bodies? Hartlett was on his knees, army cap off and a shawl of some sort wrapped around his shoulders. I only saw him do this one time, maybe the last victim—but he had his hand clenched, thumb extended. He made a motion on the man's forehead, lips and ears. It lasted just a second."

He took her hand, pulling the glove off, bringing it close to his face.

"Show me Karin—show me exactly what you think he did!"

Karin clenched her fingers, extended her thumb and touched Quinn's forehead, mouth, then both ears.

"I think that's what he did. It happened so quickly and the camera wasn't focused close-up. I . . . I could be mistaken."

165

He held her fingers to his lips, staring into her eyes.

"Unbelievable! That's called the Sacrament of Last Rights—a tradition of the Catholic Church. It is done mostly by priests but anyone can do it for the dead or dying."

Puzzled by the revelation, Quinn closed his eyes. *A Russian commander on his knees, wearing a shawl, giving the last rites of the Catholic Church to a dead hostage. It doesn't make sense.*

Squeezing Karin's hand, he asked; "What else—you said he had a shawl on, was there anything else that looked different?"

Karin was silent for a moment then, with eyes closed in forced concentration, gave as accurate a portrayal of Hartlett as her memory allowed. Quinn listened in amazement but said nothing.

"What about the Doctor?"

"Doctor LaMont thanked me for what I'd done, said to give you his thanks as well. He also had a message for you.

He wanted you to know about the Eighth Law of Prophecy."

"The Eighth Law?"

Karin pulled Quinn's head close, her warm breath on his face.

"I think I remember. It goes like this; "*If it can happen, it will; if it can't happen, it might.*"

*　　　*　　　*

At twenty-eight thousand feet, in airspace cleared of all commercial aircraft in all directions, a Navy E2C Grumman Hawkeye passed over the southern edge of the Iron Range.

Turning east over the VOR at Grand Rapids the twin-engined aircraft picked up radials for a holding pattern. Entered into one of its computers, the aircraft could circle for over fourteen hours, providing a field of radar coverage encompassing over three million cubic miles of airspace and keep track of over 600 different targets. At the moment it entered the downwind leg of the course, it was concerned with only two aircraft—an F15C "Strike Eagle," flying a similar course at 23,000 feet and a companion aircraft lower still, at 18,000 feet, both crews, veterans of over 200 sorties in Iraq.

High above the raging blizzard, the pilots, flying on instruments, could see dawn glowing gold on the eastern horizon. All three aircraft in the stack were throttled back to conserve fuel and keep the pattern tight. In the middle of the flying sandwich, the radar observer—RO—

looked over data. Coordinates were checked and double-checked. Cursors were aligned over the target area—apex of an isosceles triangle, the base of which was a north-south line running between the Engine House and Head Frame. The RO pressed a button and a pixel blinked on the screen.

"Radar acquisition complete. Target identified. Going to laser lock."

"Acknowledge. Going to laser lock," replied the pilot.

Another button was pressed and a ring appeared around the target.

"Laser lock affirmative. Going auto."

The pilot took his hand from the fly-by-wire stick. The F15 radar platform was on full automatic, laser beam coupled to the autopilot, keeping the aircraft in perfect banked orbit above a pinpoint on the ground—the fuel trucks located one hundred yards south of the Head Frame. The APC rigged with Stingers was parked close by. Another APC, rear hatch open, was being loaded with fifty-five gallon drums, the two-man team oblivious to the aircraft high above.

At Flight Level 28, Navy pilots and crew listened in silence but watched the radar screens. The show below was all Air Force.

The pilot at Flight Level 23 spoke to his counterpart below. "WHISKEY JACK to DEMON RUM, we have laser lock. Will maintain until advised mission is complete."

In the F15 at Flight Level 18, the pilot moved his gloved hands quickly over the arming toggles, flipping up safety caps. His RO watched a matching radar screen. "DEMON RUM to WHISKEY JACK, we're ready."

Redundant arming switch flipped, the pilot banked the twin-tailed fighter-bomber. Its pivot point, Sturgeon Lake, north of Hibbing. Swinging around, nose now pointing due east, the pilot flipped the last computer switch and took his hands off the stick.

Over a small body of water called Dark Lake, the aircraft nosed upwards. A laser-optic device in the nose cone of the stubby-winged rocket was locked onto the laser beam miles ahead of it. Near the top of its vertical climb, bomb and weapon separated, the Paveway II, precision-guided gravity bomb arched up and toward Tower-Soudan, the fighter describing a smooth inverted roll in the opposite direction. *"Toss"* completed at 0714 hours. The bomb curved higher and then, at the top of the parabola, lofted over, on-board optics locked onto the laser beam stretched between fuel trucks on the ground and WHISKEY JACK high above. Like a lead ball in a long bore rifle, the 2000 pound, high-explosive weapon rode the laser beam earthward. The RO in DEMON RUM watched the infrared camera for deviations from the target. Course corrections were not required.

*   *   *

Crouched in the snow, below the rock line, Quinn stared at his watch. *One Minute to go!* The wind blew away any chance of hearing the jets. He shook the snow from the bag containing dynamite, bow and radar detector. Arm wrapped around Karin, his body, shielding hers, he waited. On either side, Nick and Lubo crouched. Only the vapor of their breathing marked their presence.

Above their heads, a long, drawn-out scream pierced the dark morning air. The burst of light, sound and shock wave was followed by a blast of snow-melting heat. The sky was suddenly a psychedelic yellow-orange over the top of the Alaska Shaft. Secondary explosions quickly followed. In an instant, the fuel to sustain the operation was gone and the Stinger threat neutralized.

"Go!" Rising, lifting Karin off the rock, churning through the snow, Quinn plowed forward, knees rising, falling, the supply bag bouncing off his back. Ahead and to his left, the skeletal outline of the Head Frame was stark against the burning diésel fuel. He knew a tank was somewhere in the darkness ahead, hidden below the ironwork.

Karin stumbled and everyone piled up in four feet of snow.

The mishap saved their lives as a sustained burst of machine gun fired raked over their heads. Quinn, breath coming in strained jerks, listened as the gunner swung from left to right, short bursts now barking from the tank top. Keeping his head beneath the snow line, he whispered to Nick and Lubo, "Firing for effect—don't think he's seen us. Time to crawl!"

Suddenly, individual snowflakes turned to diamonds. The sky above, brilliantly lit. Quinn halted their forward movement. Karin, Nick and Lubo huddled together as a powerful searchlight swept above their heads. All of them heard the tank drop into gear at the same time and the clanking sound as it left its position next to the Head Frame.

168

"We're in trouble. Tanks moving. They'll spot our tracks sure as shit!"

Lubo rose to his knees, reaching for his bundle of dynamite.

"Keep going, I'll play games with this bastard!"

Quinn hesitated, Lubo was playing against long odds.

"Move Damnit or we all die!" barked the burly logger.

Arm around Karin, Nick on his heels, the three plunged through the snow, sound of the tank churning behind them.

Lungs aching from the exertion and cold, Quinn dropped into the shadow of the Dry House, its windows shot out, door blasted away. They were safe for a moment. He turned toward the tank. Lubo was nowhere in sight. Light from burning fuel cast an eerie glow over the field just covered. Their path stood out like the centerline of a highway.

Crossing their trail, the M48 swung around and lined itself up directly over their tracks, searchlight deflected to its lowest angle. Slowly, it churned toward them, Quinn motioned Karin and Nick into the Dry House. They'd have a few extra minutes at most.

Karin collapsed on the floor, Nick, gasping for breath, leveled his automatic weapon on a shattered glass window ledge, Quinn, M3 taken from the Dry Room earlier, cradled in his lap.

"Jahah, by God! Colonel, look!"

Quinn stared at the front of the tank. Lubo had thrown himself in its path. Rising up an instant before it passed over, the big man grabbed at the tow posts inside of each track. The tank lurched forward and Quinn guessed at Lubo's strategy.

"Nick, he's gonna let the tank make a plow out of him and then drop into the slot and come up behind!"

The old Finn's head shook in wonderment but held his *Suomi* level.

They saw Lubo's hands release from the posts. If the tank turned right or left, the big lumberjack was a dead man—and they were next in line. Suddenly it stopped and the turret began to traverse to the rear.

The top hatch opened and two hands reached for the M60.

"Cover me, Nick!"

Firing short, methodical bursts from his M3, Quinn saw the arms drop back into the hatch, the M60 left pointing skyward. Suddenly, a dull explosion behind the tank blew a cloud of snow into the air.

169

Bogies grinding apart, the tank jerked forward and stopped, turret swinging round again. Kneeling, dynamite in hand, Quinn struggled with a dollar lighter. It flared once and caught. On his feet, he ran forward, clambering onto the tracks, fuse sputtering. A bloody hand reached up to pull the hatch shut. As the cover came down, the dynamite went in. Quinn rolled off into the deep snow.

The ground shook. Flames licked out the hatch blown back open. Secondary explosions began. Quinn wanted out before cannon rounds started to cook off.

Quickly circling to the rear, he spotted Lubo, pushing himself up. The charge had singed his white sheet and face. He rose to his feet swaying, smiling through missing teeth. "By Goff, we diff it!"

*   *   *

The world, divided into eight geographic sectors, was being presented on the SR viewing screen, one section at time. Yoder summarized. The President noted the percent of change in traffic intercepts being reported by each station, scribbling questions on a yellow pad. He noted none reported a *decrease* in traffic—all were at two hundred percent or higher.

Yoder gave his final report ending with the comment; "The radar at Krasnoyarsk, despite our request they shut it down as a sign of good faith, is still operational and is now in phase with their facility at Pushkino outside of Moscow. My Norwegian listening post at Vadso— eighty miles from the Northern Fleet Headquarters in Severomorsk indicates imminent ship movement on a major scale."

Admiral Montgomery, responsible for keeping the Russian Fleet contained during conflict, leaned forward as did General Benedetto of the Air Force.

The CIA director pointed his pencil toward the Admiral as an afterthought. "There's been a bit of positive intelligence fallout from the LaCrosse satellite. It appears that the submarine protection tunnels at Gremikha are much bigger than we thought! Our resource at Kola reported subs not ready for sea disappeared within hours. Using night as a cover, the Russians jammed them into the tunnels to keep them safe from attack."

Yoder pushed back from the table. "Everything seems to indicate the Russians intend to slip the leash on both underseas and surface

170

vessels. Admiral Montgomery, perhaps you'd care to respond to that possibility?"

Montgomery, grim-faced, nodded and rose to his feet. "If Russia has multiple military hearts, Kola Peninsula is one of them. In addition to Fleet Headquarters in Severomorsk, the submarine base at Polyarni, there are over forty military and navy airfields on the Peninsula alone. Our inventory analysis people spot them over 500 surface combat ships, 60 of which are heavies, more than 170 submarines, 90 of which are nuclear and missile-loaded giving them a five-to-one advantage over our total NATO forces. Long range air cover is provided by Backfire and Badger bombers. General Smallzreid has confirmed that Russian army has 120,000 combat-ready troops with an additional 300 ground support aircraft under the Army Command. Very simply put, Mr. President, if this bag of snakes is let go, virtually all of the Navy's Atlantic resources and a considerable percentage of the Air Force will be consigned to a purely defensive battle with an attrition rate that boggles the mind. There are two immediate alternatives: we strike first or we activate the nuclear minefields in the GIUK Gap and tell the Russians that we've done so. If we let their ships and subs pass through that choke point unchallenged, we'd best figure on a long war . . . ."

Pointer in hand, the Chairman, face showing the rigors of the last twenty hours, sat down.

In order to be closer to the screens, the President had shifted his base of operations from the end of the conference table to the middle. He turned to the NSA. "Tom, how do you read the situation?" The President grasped the pen in both hands.

Hart took a deep breath; "We can't decode this volume but in a nut shell, I'd say the increased traffic has every Russian operation world-wide being advised to be ready for new orders and this is the critical point—respond quickly to those orders once issued."

Rolling the pen between the fingers of both hands, he looked to the Secretary of Defense. "And how do you see it?"

The Secretary nodded, "They are doing exactly what we're doing—telling our forces—stand by, stay alert and be ready to move. Like the Russians, we're not saying what the orders might be."

*   *   *

Karin re-wrapped Lubo's shoulder with a portion of her own white parka, Quinn pushed the antenna of the PRC-77 out the broken window.

"GATEKEEPER. Who's out there?"

"KHE SAHN here. Pleased to report that BRIGHTSTAR is on board. Five miles straight up!"

A new voice filled the headset. "BRIGHTSTAR'S with you GATEKEEPER and you're on open patch to the SR. We have you under full radar and infra-red coverage."

About to ask Holmes for a report, Nick poked his arm. "Jahah! Look what's happening!" He turned to see Nick pointing down at his makeshift radar detector, half covered by the supply bag. An intermittent red light blinked and when it did, a barely audible buzz could be heard. Quinn stared at his jerrybuilt device. Frequency of light and sound told him still another radar unit was operating somewhere. Suddenly he remembered the disks—still in the building. He quickly retrieved them.

"BRIGHTSTAR! GATEKEEPER here! Do you show anything moving north?"

"Affirmative, Ground, we see two bogies, in tandem, moving up on the fuel dump—or what's left of it. One target is a tish smaller—are there units smaller than tanks on the deck?"

A shiver of cold convulsed Quinn's body. He was running out of ideas and his watch told him, valuable time was being consumed and he was no closer to gaining access to the mine below.

"APC's scattered around out there—seven total."

"BRIGHTSTAR here. We'd say you're going to have a tank sweeping radar and an APC for company in approximately five minutes. Do you need assistance?"

"BRIGHTSTAR, we'll have to handle this on our own. Signing off!"

"Nick, get ready! Grab the grenade bag!" Shuffling over the floor, Nick reached Quinn's side. "We've got more problems, Nick. Another tank with radar is coming up the hill and this one's got an APC in its shadow. If I'm guessing right, they have two targets on their radar— close together—the transponder disks in my hand and that's just where

172

they're heading! Is there a way to get on top of this building—overlook the slope of the hill?"

Nick's head bobbed in the dark. "Don't know about the roof but there's a fire escape on the east wing—second floor. That be high enough?"

Quinn nodded, grabbing the bow bag. "Let's move!"

Dropping next to Karin, he patted Lubo on the good arm, gave the young woman's hand a quick squeeze. "Keep your weapons ready. Nick and I'll try to keep some distance between you and the armor."

Karin's eyes blinked under the white hood, and she squeezed Quinn's hand for a second.

Nick led the way through the dark building. The two men worked their way up two flights of stairs, down a narrow hall coming to a halt at the steel fire door. It took the two of them, pushing with their shoulders to get the iron platform clear of snow. When they did, the sound of clanking tracks crunching snow could be heard in the darkness somewhere beyond the Head Frame. The wind, eddying in the lee of the building, took their breath away.

"Maybe, Nick, we can play an old Finn trick on the tank and that'll leave the APC an easier target. Remember what you told me about *Lake Ladoga*?" As he talked, Quinn was removing the bow from the bag. He looked up, Nick was shaking his head in disbelief.

Rummaging, he found the camouflage tape. Taking a target arrow, he put it between his knees "Light Nick, hold the lamp on the shaft of the arrow, point it away from the door." Taking the small transponder disks out of his pocket, Quinn quickly wrapped one to the shaft below the tip. He repeated the process with the second disk and arrow.

"These are gonna be *our* snot-nosed Finnish Ghost patrols—our *Lake Ladoga*—that pit across the road from the Head Frame." He glanced at the radar detector, ears tuned to the audible signal. He guessed the intensity of light and sound had doubled since they left the lower unit of the Dry House—armor was closing the distance—quickly. Arrows complete, he stepped onto the balcony, staring into the white abyss. The Crusher Building was a low, dark mass directly ahead. Above, angled iron on the Head Frame. Somewhere below, the sound of tracked vehicles increased in volume. He guessed the lead vehicle was less then fifty yards away.

173

He knew the distance from Head Frame to the pit's edge was less than forty yards. The distance across he wasn't sure of." Nick, how wide is the pit on the north-south axis?"

There was a moment of silence, save for the rumble of the approaching tank. "Jahah! A stone's throw. When the younger men gots off the cage, they threw rocks for penny bets."

Quinn nodded, a hundred yards was within the range of the sixty-pound pull strength of the compound bow. Facing what he estimated was forty-five degrees on the compass, he pulled back the bow until his finger slid over the taped disk, made a silent prayer and released the arrow in parabolic flight. If he misjudged, the arrow might fall short, into the pit and out of detection range. The other danger was hitting the graywacke and granite wall, smashing the disk on impact. Optimum flight would put the arrow, miniature transponder intact, somewhere in the snow or lodged in a tree. The clanking grew nearer and suddenly it stopped. Above the wind, Quinn could barely hear the tank's engine. Smiling, he gripped Nick's arm.

"Good news—for the moment—they had to be watching one target—now we'll give them something else to think about!" Quickly he mounted the second taped arrow and released it in the same, arcing curve but thirty degrees further east.

Both men, crouching now, stared down in the direction of the tank. A sharp metallic sound told Quinn the tank commander had shifted back into gear. Soon, the sound of treads crunching snow could be heard below.

Nick, eyes used to a lifetime of staring into the dark, saw the hulking shape first, nudging Quinn's arm. His automatic rifle came up, barrel resting on the edge of the fire escape platform. Quinn released the safety of his weapon.

He wondered what the radar operator was thinking—what information was being passed up to commander and driver regarding the sudden-re-location of two fellow Russians.

Slowly, the tank came into shadowy view as it drew abreast of the Dry House, its radar antenna mounted on the turret. Quinn watched the cannon swing about fifteen degrees—then another fifteen degrees. Another smile creased his face. The cannon had been aimed at the first floor of the Dry House and now was swinging through due north. The tank paused, jerked and then slowly moved forward. As it passed beyond the edge of the Dry Room where Karin and Lubo lay huddled, the cannon split the difference between the arrows. In less than a minute the dark shape was lost in the snow but the sound continued. Quinn mentally calculated the closing speed, heart pounding, he gripped Nick's free arm. The old Finn, understanding clearly now, muttered under his breath, words coming like a small boy playing trains. "Jahah, Jahah, Jahah, Hi!"

The screech of the tank hitting the chain link fence carried over the wind, seconds later gears shifting accompanied by the grating, drawn-out sound of a massive weight being pulled by gravity over granite rocks reached their ears. Despite not being able to see, Quinn, no longer smiling, could clearly picture the tank in its death knell, like a prehistoric mammal at the tar pits and one foot was already stuck—jerking right, then left, tracks losing purchase on snow and rock and then, nose dropping as the driver frantically tried to put the machine into reverse and seconds later, plunging ten stories into the black pit.

The sound of the tank hitting the rocks far below reberverated throughout the night—a series of secondary explosions began. For the space of several minutes, Quinn was reminded of the main ammunition dump going up at Khe Sahn.

"*Sur-man-loota*" said Nick softly, wiping his nose with a gnarled hand.

Quinn turned in the old man's direction.

"Name we had for Russian tanks . . . means *'death box'*."

Inside the building, the door partially opened, they waited. Nick touched his arm, the APC, in the van of the tank, was coming into view. Suddenly, the night sky was ablaze in light. The 1.5 million candlepower search lamp illuminating the snow-filled skies. Clearing the sloping incline, the light momentarily caught them in its glare then dropped to ground level. Not sure whether the tank commander had time to warn the APC behind it before plunging into the pit, Quinn mentally prepared himself to do battle again. Radar was one thing, light another. The APC commander, more cautious than the tank driver, crept slowly down the tracks of the tank that no longer communicated.

"Water, Nick, we need enough to make ice balls," Quinn whispered to the Old Finn.

"Jahah! No water up here—buildings shut down for the winter—lines drained. I could drain my pipes for about a cup though. Your *Lake Ladoga* stunt was something else!"

Quinn laughed but was unzipping his trousers. "Quick Nick, piss into the snow that's blown inside the door."

Muttering something unintelligible, Nick obeyed and two streams of steaming urine arched into the snow mounded on the floor. Zipping up, Quinn handed Nick a grenade. "Do exactly what I tell you. Pull the pin all the way out, keep the arming bar depressed. Grab a handful of wet snow and make a hard packed snow ball around the whole thing."

"Jesus, Mister Marine—you know what you're doing?"

"I'm not sure of anything anymore, Nick, I just know we need to buy some time and get that APC or Karin and Lubo are goners!" Qinnn pulled the pin from one grenade and quickly packed the damp snow around the firing arm. Nick watched and then did the same with his grenade. Satisfied with his, Quinn reached out and carefully laid it in the snow massed outside the door. Checking Nick's he added some additional snow and it too, was put outside to freeze. Reaching for the bow for the second time, he rigged the monofilament-line device used to trail hunting arrow then pulled a T-shirt from the bottom of the bag.

Behind and to the right of the tank, the APC made a more difficult target. Its light went on and off, the commander not willing to give the enemy a sustained target. Glow of the burning fuel was rapidly diminishing.

Advancing, the huge searchlight came on, motion of the APC telling Quinn the commander was unwilling to emerge from the protection of the armored personnel carrier and was using the tracks to shift the beam. Advance and stop, advance and stop. Quinn watched and waited for the rhythm to become established. Intense light swept the area between the Crusher Building and Dry House. Despite the snow, nothing could have moved within fifty yards without being spotted by the APC. Lurching to a stop, the searchlight blinked off. Quinn pushed open the door and stepped out on the balcony. Taking aim with the bow, this time at a flatter trajectory, he waited until the APC moved ahead again. Arms beginning to ache, he held the compound bow full back, arrow sighted above and ahead of the personnel carrier. The instant it stopped again, he let the arrow go,

176

monofilament reel spinning. Line paid out, he cut it with one stroke of the K-Bar, tying it to the railing. Cutting a smaller length, he quickly fashioned a small sliding loop. The two grenades were lifted from the snow, stuffed loosely in the T-shirt and tied with a small section of monofilament. Securing bag to loop, he released the miniature satchel charge. Sliding out of sight farther than they could have thrown by hand, the grenades disappeared down the line. Quinn pursed his lips together. The searchlight flicked on.

Too far for Quinn to see, he had to rely on good fortune. The mono, shot over the APC, lodging somewhere in the snow beyond, had to momentarily drape the armored vehicle. Ice packed grenades would either explode on contact, hitting the side of the carrier with enough force to shatter the urine sheath, come to rest on the tank's surface and melt through the radiated heat—or lodge in the direct beam, quickly melt and explode.

Quinn prayed for the latter and counted the seconds. They waited, Nick at his side, M3 ready. At 22 seconds, the top of the APC lit up in sharp explosion. The huge searchlight turned blue, yellow and disappeared into darkness. Seconds later, Fifty-caliber machine gun fire blindly sprayed over the area.

"Covering fire, Nick. Take out the gunner!"

Nick demonstrated the economical technique of firing short, concentrated bursts. Pulling the pin of an M26 to within a fraction of an inch of tension, Quinn tested whether the mono still clung to the APC. The line was taut. The grenade zipped out of sight. Seconds later, an explosion near the top of the carrier, a fourth grenade hit near the base.

Nick stopped firing at Quinn's command. There was no sound save the winds. They waited five minutes but the APC remained immobile. Quinn motioned for Nick to move back. They had to keep on schedule.

*   *   *

Sporadic gunfire to the north told Quinn that Holmes—and hopefully Running Fox—were keeping the Russians in the Engine House occupied. Crouching in a corner of the Dry House, as cold inside as out, he tried to reach Holmes located somewhere in the tree-rimmed rocks. Lubo maintained a watch between them and the Head Frame, Nick, on the opposite side, peered in the direction of the Engine House. Flames fed by puddles of fuel had died down but a dull glow below the Head Frame could still be seen. Karin, huddled next to Quinn, kept an eye on the doors leading into the building. Several minutes passed before Quinn raised his former RECON leader.

"Ernie, what's happening in your area?"

"Colonel, we're putting intermittent fire into the Engine House, keeping the rounds high to prevent damage. My men are on each flank within shouting distance. Your man Running Fox is with me—came out of nowhere. I was staring into the snow and felt a tap on my shoulder. Neither of my men saw him either. BRIGHTSTAR'S on line and ready to talk."

Looking around the darkened room, Quinn saw lines of snow snaking across the floor. Wind worked its way in eddies around his small band. From the gutted building he was now in contact with Holmes on a hill a hundred yards away, with a sophisticated electronics aircraft five miles above, and through a satellite, in touch with Washington. Communications in good working order, he felt attached to the real world and the enormous military resources he knew were available. A sudden burst of cold wind reminded him the toughest job was still ahead.

"This is GATEKEEPER! Go BRIGHTSTAR!"

"Washington needs an assessment. You're patched into General Henderson. Go!"

"GATEKEEPER here, General. Compliments to the Air Force. Looks like the fuel supply's been totaled. With luck, individual tanks will be out of petrol by mid-morning."

"Sounds good, Caleb. What else?"

"Too early for a body count but I believe you can scratch two more Russians—possibly four if anyone died in the bombing. I've got a two-man patrol out trying to get to the last radar tank. Once that's shut down, we'll have isolated most of the defensive positions. LaMont is out of the picture."

The disembodied voice overhead broke in.

"BRIGHTSTAR here! We can tell you the Ground Surveillance Radar *is* running and tanks now appear to be communicating with field radios. Do you want the radar tank taken out?"

178

About to say yes, Quinn remembered Sergeant Maki and Wally. *They must have gotten between the tanks and cut the wires. A smart bomb would be the easy way, but it could also kill his men on patrol!*

"Not yet. I've got my men on the ground. I can't risk their lives."

"Be advised GSR tank appears to be moving north."

Quinn paused, forcing a mental picture of the battlefield to form in his mind. "That means it's coming up the hill. They must want it to replace the tank we just took out!"

"We have a fix on you and your radio base. The tank seems to be heading directly to the location you're in now, approximately five hundred yards away and moving at five to seven miles per hour. Terrain must be tough or the snow is making it difficult going."

"Thanks for the information. Stand by and let me talk with Captain Holmes."

Quickly changing frequencies, he raised Holmes.

"Ernie, we've got to move out. Is Running Fox nearby? Good! In three minutes we move to the Cable House from both sides. If we don't have our hands on those power levers, we don't get into the Orchard. Running Fox takes the northern entrance, you the west and I'll come in under the cables. Nick's with me so that's four sources of firepower. Expect two Russians, maybe more."

Nick was at Quinn's side, gun on his hip.

"Lubo, you cover our backside."

"Karin, take Nick's place. If anybody slips by they'll try to make it to the Head Frame. You know what to do?"

She nodded, her face hidden in the folds of the white parka.

Pushing open the door leading out of the Dry House, Quinn looked straight into a burst of snow gusting between the two buildings. He pulled off his trigger finger glove, yanked the hood over his head and stepped into the deep snow.

"Jahah! It's *bielaja smjert* time. *White death* Colonel, that's what the Russians called us."

*   *   *

Quinn leaned against the dark building. Overhead, he could make out the thick cables rising at an angle out of the Engine House. Jamming his bare hand under the sheet and inside the camouflage hunting coat, seeking the warmth of his armpit before the assault.

"Lights, Nick, where are they?"

"A small one, inside the door, maybe 75 watts. Main lights for the room are about ten feet further in, on a small metal platform to your immediate left. Four, maybe six switch boxes. Gots to throw'em all."

Quinn tried the door know, locked. Twenty seconds remained. Overhead, black cables disappeared into the direction of the Head Frame. "I'll blow the door, Nick. You come in behind me and get to the light panel. You ready for this white death stuff?"

"Jahah!"

Backing away, Quinn fired a burst into the door.

Steel-jacketed rounds against iron created an unholy sound that reberverated between the brick buildings. Snapping the clip out, jamming in a second, he pulled the door. Nothing happened.

Bursts of gunfire erupted from inside the building. Quinn lost track of the direction. He fired another short burst at the lock and gave it a kick. It opened, rounds from inside ricocheting off the top. Dropping to his knees, he stuck the nose of the M3 inside, squeezing off the remaining clip, reloaded and tapped Nick on the shoulder.

Bent over, the old Finn scrambled into the dark. Shattered glass rained down from windows high off the floor, whining bullets slammed off brickwork, creating a crescendo of numbing sound. Suddenly, a beam of light appeared and moved along the wall where Nick was maneuvering. A single shot rang out and the beam flipped toward the ceiling then bounced down metal steps, flickering crazily as it jangled downward. An instant later, the large room was bathed in light. Quinn blinked and dropped into a crouch, the third clip jammed home. Quinn shouted toward the operator's station directly in front of the winch drum. From under the iron grid work, Jimmie Running Fox materialized.

Holmes, near the small wooden door on the western side of the building, rose from his position, a dead Russian sprawled at his feet. With two more Russians out, they controlled, at least partially, the mechanism providing access to the mine itself. Suddenly, a cannon thundered, the direction and volume of sound causing Quinn a sinking feeling. He knew the Dry House had come under fire. Karin and Lubo were no match for a tank.

Raising his weapon again, he signaled Holmes and Running Fox to gather in the corner of the Engine House. Dropping to his knees, Quinn pulled the radio out from under his parka, the light indicating a message was on hold, burned a dim yellow. Wearily he composed his thoughts. *If the tank had sliced up the hill it must have headed toward the Dry House. Now we have to disable it before we get access to the Head Frame.*

*   *   *

The mood in the Situation Room was somber, the silence, intense. Twelve minutes past the hour, every member was present and everyone had read and reread the text of the Russian reply.

Pushing back from the conference table, leg throbbing, the President felt the need to do something other than sit and stare at this list of denials followed by a list of demands. He stood under the screen facing his advisers.

"We catch the Russians red-handed and they come back with this bullshit! They deny any knowledge or responsibility for persons claiming to be Russian by birth and members of the *Voiska Spetsialnovo Nasnacheniya*. And—they accuse us of fabricating a *'cause celebre'* in order to strengthen our military position on a worldwide basis. As our moves have been *'unilateral in nature'*, they feel free to take counter-matching moves on a case-by-case basis. The primary step being to leave the Krasnoyarsk Radar running in the defense of their country."

Head shaking, the President tugged at the folds of loose skin around his neck. "To further complicate matters, he admits in a postscript, the names we provided match those known by them to have been members of the SPETSNAZ who did serve in Afghanistan. And all four came from the Ukraine."

Perplexed, the President turned to face his staff. "He also contends the four are designated as Missing-in-Action and presumed dead due to the *Mujahedin's* no-prisoner policy." He raised his hand, pointing over his shoulder. "Our friend in the Green Room also chides us for our continued assistance to the *Mujahedin* . . . ."

Admiral Montgomery stood up. Jacket off, tie removed, cuffs rolled up. "Mr. President, speaking on behalf of the military, we believe it imperative we go to full nuclear alert status. Failure to do so is to grant the Russians, on a case-by-case basis, five to twenty-minutes advantage on each and every front."

Leaning on the table, eyes locking onto those of the Chairman, the President spoke softly but firmly. "I told the American people—and the world—I would not be a *First Strike* president! I said it then, and I'm repeating it now on the firing line!" Fist slamming down, the closest coffee cup bounded off its saucer, a brown stream racing toward Tom Hart's lap. When his eyes met the President's, both men knew it was time to call for a break.

*　　　*　　　*

Eli Ben-Yaacov poured tea over four cubes of sugar for the Prime Minister and fresh orange juice over ice for himself. Despite disparate dress and appearance, Prime Minister, pale of complexion, in a black-striped, double-breasted suit, the head of the secret service, deeply tanned, in khakis, the two men had compatible minds.

Together, they studied the text and facsimile photos bounced from Tangier to the Ben-Gurion down-link. "Not good, Ben-Yaacov, not good at all. The Americans seem justifiably provoked and the Russians blithely deny any wrongdoing. Meanwhile, both gird their loins for battle."

Ben-Yaacov nodded. "The last time the Hot Line was used was at the onset of the 100-Hour War. I agree, Moshe—this is on the high end of the diplomatic Richter scale."

Pulling the photo file closer, finger tapping the pictures of the lieutenant and corporal captured in Washington, the Prime Minister studied the photo. Whip marks, referenced in the text, while not pronounced because of the method of photo transmission, were nevertheless discernable.

"Why, Ben-Yaacov? Why would Russians bear markings like these?"

The spymaster laughed, shrugging his shoulders. "Perhaps I should go flying again Moshe. I do my best thinking when the airspeed indicator is beyond Mach One."

Snorting in agreement, the rotund Prime Minister poured his own tea. "I listened to the conversation over your radio system, Ben-Yaacov. You take great delight in making fools of the Jordanians. The entire

182

Israeli Air Force thinks of you as a hero out there in your *Kfir*. Testing their response time, you say. I say that you are testing yourself!"

Ben Yaacov's curly black head bobbed with delight. "Two birds, Moshe, I always look for two birds."

Eyes growing serious, the Prime Minister tapped the message. "Well, Mr. Two Birds, make something of this. I see trouble on the horizon."

Pushing the orange juice glass to the side, Ben-Yaacov studied the photos of the prisoners, disks, fragments of a cyanide pill and the list of names provided by the Americans. Several moments passed before the *baleboss* looked up; "Moshe, if the Russians are bluffing and these people with stripes on their backs do report to the KGB or GRU, that issue is one thing. But note that the Russians deny—but do *not* unequivocally distance themselves from these people. Perhaps that is where the truth is?"

Head shaking from side-to-side, the Prime Minister appeared baffled. "Your logic escapes me, Ben-Yaacov. I see no truth here. None whatsoever."

Ben-Yaacov raised his hand, a smile on this face. "I apologize Moshe, my mouth is like the *Kfir* itself; it can move faster than the human brain. Allow me. If the Russians *did* authorize this mission, then one can assume that they felt the juice worthy of the squeeze and are now simply buying time. If you accept their denial at face value but consider the American evidence, the failure of the Russians to totally distance themselves from the captives could mean they themselves don't know who they are."

"How can that be? Surely the Russians are in control of their own intelligence apparat?"

Eli Ben-Yaacov laughed heartily. "Ah Moshe, surely you jest. In the prior coalition here, one individual took it upon himself to recruit American Naval personnel for Israeli intelligence purposes. We disavowed his actions and apologized to our U.S. counterpart. An embarrassing episode but it actually happened."

"Granted, that did happen. But tell me, what does *this* government do with the information? This little Hebrew who happens to be Prime Minister, believes the giants have drawn a line in the dirt and bad things can happen. Remember what you said earlier?"

"I do Moshe. It bears the closest watching, and we shall soon confer with Rabbi Rachman. He is on his way here now."

"Rachman agreed to leave his beloved mountain top? I do not believe!" The Prime Minister was incredulous that the respected scholar could be enticed away from his isolated retreat and olive trees.

"Moshe, it took me years but I finally discovered his Achilles Heal."

"Rabbi Rachman is not mortal like you and me, Ben-Yaacov."

Ben-Yaacov smiled at the Prime Minister's description. "He asked me why I loved to fly and I told him—seeing Israel from the eye of the hawk—the beauty of the landscape, flowering gardens, the rugged grandeur of the Negev at dawn—and I told him of saluting the dead warriors of Masada with the wings of my *Kfir*. It was then that he expressed an interest in sharing this joy with me. He is coming here now, in my helicopter—*'on eagles wings'* as he puts it."

*   *   *

Hart sat perched on the desk, the President in the chair, leg elevated on the open drawer. Closeted again in the small office adjacent to the Situation Room, he sought private counsel with this National Security Adviser.

"Hurts?"

"I swear to God Tom, this knee's a political barometer—the shittier the situation, the greater the pain." He looked up at his NSA, "What do we do now? Joint Chiefs, Yoder and even the Secretary of State are like piranha and there's blood upstream. And you—if we went back in and voted now, which way would you go?"

Hart fingered the tip of his tie, eyes down cast, silent for a moment. When he lifted his head and looked into the eyes of the President, into depths of sadness he had never seen before. "If we voted now, on the basis of information available, I'd go with the Joint Chiefs. We're exposed everywhere to Russian first strike capability. You've seen the battleground assessment. If they take the first shot—the score is 7-0 in their favor and we start our game deep in our own end zone. Going to Nuclear Alert, we at least have a fighting chance."

Rubbing his knee, the President studied him. "And Tom, we might just provoke the Russians into pressing the button, a self fulfilling prophecy. I don't know if Trishenko has a bag man outside his door but

184

I'm hesitant to have mine make that move from hallway to conference room."

The NSA looked at his watch. "We've got just ten minutes to send the next text. Yoder, Montgomery and Weller should have a draft ready now. You ready to go back in?"

"Give me a couple of minutes to collect my thoughts," said the President shaking his head.

Head tilted back, fingers steepled, the President stared at the ceiling. Slowly, he lifted his leg off the drawer with both hands, stood up, and slipped on his loafers. The moment he stepped into the hallway, the army major rose from his chair, coming to rigid attention, eyes riveted on the wall ahead. The face was unfamiliar to the President. He walked toward the door leading into the SR and then, on sudden impulse, took the extra step placing him directly in front of the man with the nuclear strike codes. The President could see already stiff shoulders tighten up. Four inches taller, the man's eyes were hidden by the polished visor of his dress cap. The black plastic name tag identified him as R.A. Hughes.

"Stand easy, Major. Take your cap off."

"Sir!"

"Where's home, Major Hughes?"

"Salt Lake City, Sir!"

"Relax Major, remember, I said stand easy.

"Salt Lake. Does that make you a Mormon?"

"Yes Sir!"

"Career military?"

"Yes Sir!"

"This must be your first time on rotation?"

"Yes Sir!"

"You had to volunteer for it. What goes through your mind when that bag is wired to your wrist?"

The young officer didn't answer, seeming instead to read the President's face—asking silently if his response was truly being sought—the President nodded.

"I pray, Sir, that I or anyone who carries it for you never has to open it for any reason."

"Even if the Russians fire first Major?"

185

Hazel eyes studied the Commander-in-Chief for a moment. "I was taught, Sir, to turn the other cheek."

Left hand locked under the elbow of the right, right hand stroking his chin, the President stared at the young officer. "I see a wedding band. Any children?"

Major Hughes brightened. "Twins. Two boys, age ten, Sir!"

"I understand that territory pretty well. Back to raising them, however. If one twin hits the other twice, would you recommend turning the other turn the cheek twice?"

"No Sir! I teach them the cheek needs be turned but one time."

The President put his hand on the officer's shoulder. "Tell your boys I want to meet them—make sure of that Major—at the White House Christmas Party."

"I'd consider it an honor, Mr. President!" The Major beamed.

Turning, the President left the officer with the Doomsday codes in the hallway.

* * *

The sharp, barking staccato of a machine gun filled the air. Huddled in the lee of the Engine House, the Dry House, a gray mass ahead of them, Quinn, Holmes, Nick and Running Fox tried to assess the problem. There was no response fire coming from within the Dry House. Gut churning at the thought of Karin and Lubo dead at the hands of the Russians, Quinn grabbed the radio.

"BRIGHTSTAR, GATEKEEPER, here. More problems! Is everybody listening? General Henderson, we own the Engine House but it's an empty victory. The GSR tank is now between us and the Head Frame. We're too bunched up to ask for laser bombs at this point. We may have lost two more of our people."

Quinn felt a hand on his forearm. It was Running Fox. "The firing stopped on the other side of the Dry House, I'll do some quick re-con."

Pointing to the radio in the Captain's hand, Quinn nodded. In an instant, the Ojibwa faded into the darkness.

"Caleb, time is getting short. Can you get into the mine? Can you stop the computers?"

"At this moment the answer is *no* on both counts."

"Then consider this, Caleb. You pull back and the Air Force levels the entire area."

186

Quinn stared at the radio in disbelief. Nick swore in Finnish under his breath and whispered in his ear. Quinn's head moved in agreement. "Absolutely no way general! If the shaft collapsed it would take weeks to get to the bottom level. The people down there would die of starvation in the dark. That's not an alternative."

"Colonel, Running Fox on the line. Lubo's about had it, took one in the gut and played dead. Two Russians stormed the Dry Room and grabbed Karin. They thought she was one of them. She had the disk in her pocket and it showed up on their radar. Lubo said she meant to tell you she brought it along—LaMont suggested it—but she forgot. Lubo thinks she's going to be brought down into the mine as another hostage."

"Colonel!"

Quinn turned, Captain Holmes pointed through the open door leading back into the Engine House. A flashlight beam filtered through the air.

"Gotta be Russians Colonel. If they're going down, they gotta have the Engine House!"

"Nick!"

"Jahah! I'm here Colonel."

"You said it was possible to ride down the cage on top. Is there a way to get into the mine without going through at the Head Frame?"

The old Finn started to shake his head then stopped. "Yah! there is! We goes back to the Alaska—snake down to Level One and straight back to the Head Frame. We'd be a level lower but could crawl up the safety ladder to the top of the cage. By gully it could be done!"

"Ernie take a radio, go back around the Engine House. Find your men and stand by the rig. We'll give the Russians access to the Engine House, but not until we're ready. Fir' into the building to keep them honest! I'll radio when you can cease-fire. Nick and I'll pick up Running Fox and move back to the Alaska Shaft. We're going in on the same run that brings Karin down!"

He pressed the radio. "BRIGHTSTAR, Jocko, we're going for the mine shaft. Take out the satellites and when the weather breaks, have the fire bombers and chopper drop water on the tanks. Without fuel and with the temperature dropping, we can freeze them—water in the

187

hydraulics could keep the turrets locked, ice in the barrels makes firing them downright hazardous to their health!"

"Sounds crazy as hell, Caleb, but you've got it! I've gotta tell you, though, that taking out the Russian satellites just might send up the balloon."

"Jocko, pass this on to Mumford. The Russian commander here might also be a priest! And somebody should look into the whip marks. Could they be self-inflicted?"

There was a sudden pause.

"This is ROCKAWAY, remember me, I'm still patched to your ass. What's this about a priest?"

Forgetting momentarily, the CIA analyst was part of the team, Quinn smiled in the driving snow. "It's not much to go on, Mel, but Karin saw him give what I'd describe as last rites of the Catholic Church to one of the hostages. I didn't see it but she showed me exactly how he did it—thumb over the forehead, lips and ears. He was also wearing a shawl of some sort." He quickly explained the other details remembered by Karin. "I gotta go, Mel! There's daylight out there! We gotta move!"

"Good luck Ground," came down from the AWAC.

"Semper Fi, Caleb," from Henderson in the Situation Room.

"ROCKAWAY here—take care!" from Mumford at Langley.

$$* \qquad * \qquad *$$

"We try one more time!" The President looked right, then left, sentiment in the room was for moving to Nuclear Alert and advising the Russians over the Hot Line that the non-nuclear portion of the mine fields in the Greenland-Iceland-United Kingdom sector would be activated thirty minutes past the hour of receipt. But the President prevailed, and once more the move to Nuclear Alert Status was temporarily delayed.

In addition to the last-minute intelligence about the SPETSNAZ commander, notification was given to the Russians that the United States Air Force would, barring last minute conditions on the ground at Tower-Soudan, take out the Russian weather orbiters known as Cosmos 948 and 1222. Rather than admit that the Russians still had the upper hand in Northern Minnesota, these "*last minute conditions*" were not described.

Less than two minutes to the hour, the President, on a sudden impulse, added a personal note to Trishenko. He mentioned an unplanned conversation that had taken place on a foot path at Camp David, a spot where Prime Ministers Begin of Israel and Sadat of Egypt met by chance while on a late afternoon stroll in October of 1978. In the dappled sunlight, the President told the Russian leader, that fateful meeting, as much as anything made the Israeli-Egyptian Peace Treaty possible. The last line of the President's short missive; "Perhaps Yuri, history can repeat itself in the same way, same context . . ."

*

*   *

Leaving orders with Ernie to have Lubo picked up as soon as possible, they prepared for the next move. Miners helmets and long-handled flashlights were taken from the wall. Running Fox came up with the idea allowing them quick access back to the Alaska. Slipping due north fifty yards, they made contact with the link fence that kept tourists from pitching into the pit that was now a burial ground for an M48 tank. Boosting Nick over, Running Fox assisted Quinn and then clambered over alone. Wind had scoured the rock clear of snow and they were able to move at a fast crouch past the tank guarding the entrance. Holding onto the fence, they kept from slipping over the edge. The charred ruins of a tank smoldered below.

Intermittent gunfire broke out behind them as Captain Holmes began suppression fire into the Engine Room.

In minutes they were back below the lip of the Alaska Shaft. Nick leading the way at a crouch, mindful of his earlier experience, went headfirst over the ledge landing squarely on his grandson and Sergeant Maki huddled in the crevice.

"Jahah! What we got here?"

The two men embraced in the dark. Quinn and Running Fox slid down the black rock.

Maki's left arm hung loosely at his side, a dark stain showing.

"Sorry we didn't eliminate the GSR Colonel—couldn't get close enough to get a shot. We worked our way into the ring, cut wires between four more tanks but headed back here when the GSR started up the hill. Did pick up two more radios though." The sergeant patted one slung over his shoulder, Wally, smiling, held up the other.

"Good, real good!" Quinn took young Saatela's radio, selected the channel. An opaque yellow light went on.

"Channel 23 Sergeant, when you talk your code is GROUND BAKER. You'll be heard by Captain Holmes on the hill behind the Engine House. There's an AWAC on station directly above—guarding all channels. That's our fire support command post and simultaneous relay to Washington." He laughed. "And, unlike any war you may have heard about, you'll be talking to the Situation Room in the White House.

Nick, you check out the passage way to Level One. Take Wally. And when you have it located, send him back. I want to give last minute instructions to what few men we've got out there."

Wally and his Grandfather pulled the wooden cover off the entrance to the Alaska, dropping quickly out of sight.

Maki coughed. "You got a few more men, Colonel."

Quinn turned to face the sergeant.

"We ran into another tank commander and three privates. They've got a radio and are located on the southwestern perimeter, a construction site on the edge of the mine. They want part of the action. I told them to guard Channel 16."

Quinn balled one fist in salute. "That's great. I couldn't ask for help at a better location. We can at least fire into the rings!"

He tapped Maki's radio. "Get that squad on the horn and give them a description of what's happening. Tell them to look for Russians coming out of the tanks—or trying to make it out of the ring toward the Head Frame. I'm guessing that's where they'll make their stand." About to turn, he felt Maki's hand on his forearm.

"Colonel—something I've got to ask." The sergeant pulled his white hood back. "You're not the man Karin is seeing?" Maki's voice was hesitant. "Her husband's my brother . . . ."

Quinn laughed softly. "I don't think it qualifies along those lines. Karin's been my once-a-week housekeeper this summer. Yesterday, I invited her to come up here while I did a small favor for a friend in Washington. That favor turned into this and guess who got sucked in? She's a hell of lady, Sergeant and . . ." He suddenly realized Maki couldn't know what happened.

"And . . . ?"

190

He looked at Karin's brother-in-law for a moment. "She was taken by the Russians Sergeant Maki. She's about one hundred yards away and will be going down the shaft as soon as we let the Russians take control of the Engine House." Quinn held back the possibility she wasn't at the Head Frame—that someone had decided against another hostage. There was a long silence as they stared at one another. Running Fox, impassive, sat with his back to the rock, watching the two men.

"I'd like to volunteer to go down into the mine, Colonel."

A burst of gunfire to the west told Quinn that the next act was about to begin. He looked Maki in the eyes. "I know it's important, but you're wounded. More important, you understand what's going on up here. I've got experienced troopers in three key places now—all radio connected, and that's too important to give up. You stay here. We'll do our best to see that Karin comes out alright." Quinn sensed the anguish in the man's eyes. He gripped the man's good shoulder. "Sergeant, when this is over and done, someone should be around to tell her what happened. Let that someone be you!"

Wally's head appeared above the shaft entrance. "We got an opening Colonel. Grandpa found it. We're ready to go!"

Quinn turned to Maki. "Get on the horn, Soldier. Your squad needs orders!"

Sergeant Maki pulled the hood up with his good hand and saluted.

Turning to Wally and Running Fox, Quinn issued quick orders. "Get into the shaft. Tell Nick to start working his way forward. Wally, Running Fox—stay back at least twenty yards in case he runs into trouble. I don't want everybody bunched up. I'll be in the shaft in two minutes!"

Saatela dropped from view as Running Fox slipped into the opening.

Quinn pulled up the antenna of the radio.

"BRIGHTSTAR, Ernie—GATEKEEPER here."

"Captain KHE SAHN here, ears on."

Quinn smiled. "Good. We've got more Guards, four—men on the southwest perimeter, Ernie. Sergeant Maki will maintain his position on the eastern flank. His call sign—GROUND BAKER. All we can do on the surface is keep the Russians occupied. They have the fire power but we can limit their mobility."

"BRIGHTSTAR here. Infrared indicates four of the twelve tanks and three APC's showing signs of heat decay. Others still showing hot. GSR is still active and is now located at the Head Frame."

"BRIGHTSTAR and KHE SAHN, there's still the question of the primary Command Track. Whoever spots it, get on the horn. Holmes is in charge once it's been pinpointed. Give BRIGHTSTAR the coordinates and let the Air Force kick ass!"

"Understand Colonel, Command Track to be terminated on discovery!"

Suddenly, Quinn was pressed to the rock by the force of Maki's body. A mortar round whistled overhead and hit beyond the far edge of the Alaska Shaft. Pushing up, he keyed the field radio. "They're listening from somewhere! Break off! Break off!" On his knees, Quinn pushed the antenna down. "Sergeant, haul ass out of here. Keep your eyes open but head down. You know what they got out there!"

A second round screamed in but fell short. Both men knew bracketing rounds had been fired. The next one could easily drop into their crevice. Maki faded into the gray dawn.

Pulling the white, makeshift parka around his body, Quinn slipped into the abandoned entrance to the Alaska. For one brief moment, he looked up at the Norway pines, tips bending in the blowing snow.

A gust of cold wind swirled in the small rock pocket. Taking one last look at the sodden dawn breaking around him, he lowered the bag containing grenades, dynamite and extra ammunition clips to his feet and slipped into the dark chamber flecked with greenstone and graywacke. Ten feet into the shaft opening, he heard a heavy explosion behind him and felt the concussion. Turning, he saw rocks tumbling into the shaft. Shaking from cold and tension, he realized the last mortar round was a direct hit on the Alaska opening, that retreat was impossible.

*   *   *

The Russian President read the text with mixed emotions. Activating non-nuclear mines in the GIUK was an overt act of war. Any Russian ship passing through was at extreme peril. Taking out the weather satellites over North America *might* be justifiable—if it could be

192

proven they carried nuclear weapons. Manarov was conferring directly with the Cosmonaut Command at *Baikonur* to verify the status of the two satellites, Korovkin, implications of mine activation. Both men would be back in the Green Room within ten minutes. Should the troika make a presentation to the waiting politburo under present circumstances, he knew Korovkin would rule the *STAVKA*.

From his vantage-point, he stared out into the lightly falling snow, the bright red star of *Spasskaya*, looming above. Beyond the cobbled streets, he could make out the onion domes of St. Basil's Cathedral, the church, built by Ivan the Terrible to commemorate his victory over the Tartars. He could see the gray *Lobnoye Mesto*—where citizens of Moscow heard Tsarist proclamations read or—watched heads roll. His Country was undergoing tumultuous change. *Only a civilian government can guide us through—if the military takes over—we return to the Cold War and no progress.* He brought the text up, adjusting his reading glasses and read it again, concentrating on the short paragraph at the end.

Closing his eyes, Yuri Trishenko remembered the day and the hour he stood with the President of the United States, under the linden trees in the Catoctin Mountains of Maryland. He willed himself to communicate directly with the tall man who had admitted to him, *"at heart, he was just a small town boy."*

*　　　*　　　*

Rabbi Rachman, scholar, historian and holocaust survivor, stared at the text of the intercepted Hot Line messages. Placing it on the glass-topped coffee table, he leaned back in the chair, stroking the pointed graybeard that accentuated his gaunt, narrow face and deep set eyes. More than once his appearance had been likened to that of Lincoln. His gaze shifted from Prime Minister to Spy Chief.

"And *what* are the implications for Israel?"

The Prime Minister motioned to Ben-Yaacov.

"I understand your reticence at reading mail marked, "Eyes Only—President," Rabbi, but consider this—if we did not have such prior knowledge, Israel would be in flames hours after an outbreak of hostilities between the United States and Russia. Those of us lucky enough to be in an underground bunker like this, might survive only to die in a *jihad*—or we might be gassed to death down here."

193

The Rabbi came out of his chair, arms coming to rest on his shiny black trousers.

"That is a serious statement Eli Ben-Yaacov! How much is fact? How much hysterical hyperbole used to justify a pre-emptive attack on our neighbors?"

Putting his teacup down, the Prime Minister leaned forward in his chair. "Rabbi Rachman, you have chided this and other governments for aggressiveness. The Defense Minister is meeting with army, air and naval leaders at this very moment, preparing for war. If Yitzhak could be here, he would tell you what Ben-Yaacov has been authorized to tell you now."

He nodded. Ben-Yaacov clasped his hands together. "It has been known for some time, Rabbi Rachman, that should hostilities begin between the Superpowers, a tacit agreement has been struck between Russia and the Arab World. Syria leading the charge will attack Israel. We have copies of their battle plan. The most plausible situation would look like this." Ben Yaacov re-arranged the papers on the glass table, creating a rough outline of Israel. "Syria would be conducting maneuvers near Damascus. Suddenly, 200,000 soldiers and 2,000 tanks would wheel and turn on the Golan Heights. Syrian, Jordanian and PLO commandos are dropped by helicopter in Galilee, blowing bridges, roads, power lines and stations. Syrian and Jordanian jets, only five minutes away, would destroy any available target in the area. Many would be shot down—but not before they destroyed pre-positioned supplies needed for our reserve forces. While this is going on, this command post, Tel Aviv, Haifa and other major cities would come under attack by missiles—then comes the next phase—from Iraq— missiles supposedly given up after the 100-Hour War—and containing chemical warheads made in Libya with German technology. We have Patriot missiles but nowhere near enough to stop them from coming in from so many different launch-sites—so many in number. You know that even as the 100-Hour War was going on, Syria was buying SCUD-type missiles from China—twice the range of those used by Iraq to attack us. Certainly they aren't needed to attack Iraq?"

Beard grasped in hand, Rachman's head nodded in silent understanding.

194

"Our people would be in the open, exposed. The entire Arab world would likely join in the blood bath while our greatest ally, the United States is fighting for its own survival. The *'Jewish Question'* as posited by our enemies, would be answered. Israel would cease to exist. Period! The PLO would be given this land after Syria buffered its borders. Egypt would reclaim lands won by us in prior wars and held as Military Territories. The goal would be to kill as many Jews as possible during the conflict itself and let the PLO deal with survivors as it saw fit."

Rabbi Rachman straightened in the chair, eyes closing, arms crossing his chest as if the room had suddenly turned cold. His upper torso rocked backed and forth, knees tightly pressed together. To Eli Ben-Yaacov, it reminded him of the hundreds of black clad rabbis who daily mourned in a similar fashion at the Wailing Wall. Several long moments passed before the Rabbi's eyes opened.

"And you, Mr. Prime Minister, do you agree with Ben-Yaacov's assessment?"

"I have seen the documentation, heard first-hand from operatives who participated in these secret planning sessions. Yes. The plan is reality. It makes sense that the Russian would want Israel fully occupied while it fought the U.S. Now that hostilities between Iraq and the coalition have ceased, the combined powers of the Arab countries involved would be thrown against our defenses as well. This land would be a vast charnel house."

Rising, Rabbi Rachman turned away from Ben-Yaacov and Prime Minister. He contemplated the panoramic view of the Negev papered against the far wall to mitigate the fact they were deep under ground. The two men saw his hands entwined, thumbs rotating about each other. "Eli Ben-Yaacov. Do I gaze upon dawn or dusk on the Negev?"

"Dawn," answered Ben-Yaacov quietly. "You are two hundred feet off the desert floor, flying northeast between rock rifts. One is safe there from land-based radar in Jordan but . . . ."

The Rabbi turned. "But what Ben-Yaacov?"

"One is never out of sight of their airborne radar."

"We live in a fish bowl?"

The Prime Minister nodded. "A small, fragile bowl, Rabbi Rachman, easily smashed."

Ben-Yaacov added in a melancholy voice, "Others would plant, raise and pick the olives . . . ."

Lowering his thin body into the chair, arms again on the knees, the Rabbi placed his hands, palm down on the glass tabletop. "What needs to be done?"

Ben-Yaacov tapped the glass beyond the paper boundaries created earlier. "There are no maneuvers currently underway in Syria—just one of several puzzling aspects. Annual maneuvers were completed three weeks ago. However, communications between Arab nations and Russia have increased in the past ten hours and there is talk of re-scheduling joint military activities between Syria and Jordan. Egypt seems reticent to become involved."

"That is good news—is it not?" asked the Rabbi.

Ben-Yaacov nodded. "For the moment." He put his pen on the intercepts. "If there is any hope, Rabbi, it seems to rest within these messages."

The Rabbi reached for the documents. "So—there is a hidden message here?" He pulled his wire-rimmed reading glasses from his vest pocket and scanned the documents. When he put them down, he was steely-eyed, his tone stern, admonishing. "You, Eli Ben-Yaacov, have you sent these men into the American laboratory to steal these secrets? Have you inserted your uncircumcised member in a place where it is now stuck—and you wish me to assist in its extrication?" He shook the text under the Spymaster's nose.

Ben-Yaacov laughed and took the facsimile, placing it on the table. "Not guilty. This is not an Israeli operation. Both the Prime Minister and I give you our word on that. You have, however, touched on a distinct possibility—'an *extrication*' as you just put it."

"So?"

"So, Rabbi Rachman, studying these purloined texts carefully would lead one to two possible conclusions—the Russians attempted to avail themselves of American secrets and were caught—or another party, masquerading as Russians have put this effort together and is not concerned that it could lead to nuclear war." The Spymaster paused and then added as an afterthought, " . . . perhaps even desires it."

Rabbi Rachman picked up the texts, scanning them again. "I see clearly that the Americans have a grievous problem on their hands and

I see just as clearly that the Russians deny any wrongdoing. Most clearly of all, I see Newton's Law at work—for every motion one—an equal and opposite motion. And, yes, Eli Ben-Yaacov, I agree that Israel is between anvil and hammer. But what course of action other than preparing for war is there?"

"Prevent it!" blurted out the Prime Minister in exasperation.

Ben-Yaacov nodded in agreement. "In the intelligence game, it is a plus to be able to hedge one's bet. If the Russians are lying, virtually nothing can be done but prepare for the second holocaust. If there is another party involved, however, we must do what we can to expose them before either President steps over the line of no return!"

"What evidence—Prime Minister—Ben Yaacov—do we have that a third party might possibly be involved?"

Ben-Yaacov reached to the floor, picking up the pictures that had been separated from the text.

Spreading them out before the rabbi, head down, Ben-Yaacov did not see the Rachman's thin face grow pale, jaw muscles suddenly harden. The Prime Minister, to his right, saw the thin fingers tremble slightly but made nothing of it.

Looking up, the *baleboss* continued. "The perpetrators of this act—are identified as SPETSNAZ by the Americans—but claimed by the Russians to have died in Afghanistan. And they all had unusual attributes. Consider that four soldiers, three men, one woman, all have the mark of the lash—not a one-time event, but repeated to the point that the welts seemed to be of a long-term nature. And now we have this." He pointed toward the circled passage below the facsimile photo and read the caption containing a third party observation passed on by an American officer "*involved in the action.*" He handed it to the Rabbi. Adjusting his glasses, Rabbi Rachman read the short text silently, holding it tightly between his hands. The paddle fan in the center of the room, its huge blades turning slowly, made whisping sounds as it moved the air about the subterranean command post.

Finally, the Rabbi put the paper down on the tabletop. "I better understand your concerns Eli Ben-Yaacov, Mr. Prime Minister. Guilty or not, this country stands implicated by an observer who missed nothing. The Catholic blessing of the dying is one thing, but the reference to yarmulke, tallith and tefillin, with phylacteries wrapped around the forearms is quite another."

Rabbi Rachman rose, unsteady for a brief moment but placing a hand on his chair faced away from the two men, his gaze coming to rest on the paper desert.

*   *   *

"Mr. President, Baikonur reports that Kosmos 948 and 1222 are low-pass orbiters designed for imaging of weather in the Polar Regions. The only unique feature between them is on Unit 1222, an infrared frost detector. This allows our agricultural planners to determine with some precision when wheat planting should commence."

Marshall Korovkin snorted. "We have that advantage and still buy wheat from the Americans?"

Manarov glanced at the Field Marshall now in a full dress uniform, and continued; "If the Americans take out these satellites, they take out nothing of military or strategic value. Why they make this an issue that could lead to war is beyond me."

Trishenko looked across the table to the Field Marshall.

"Could *we* take those satellites out?"

Both Field Marshall and KGB chief looked at the Russian President as if thunderstruck. Korovkin was the first to respond.

"That's absurd! Why would we do that?" he roared.

"And why Korovkin—would we not?" asked the Russian President pleasantly, "especially if it pacified the Americans? General Manarov said they have no particular military value." Yuri Trishenko kept his eyes locked on those of the Marshall.

Korovkin's face turning dark seemed at a loss for words.

The Russian President persisted. "Tell me Field Marshall Korovkin, can we take those satellites out like . . ." He snapped his fingers loudly.

On the defensive, the Field Marshall looked at Manarov who quickly glanced down at the green tabletop. Receiving no support from the KGB officer, Korovkin sullenly responded; "Da! We could take them out—like that!" In a mocking manner, he repeated Trishenko's finger snapping motion.

"How Field Marshall?"

198

"*Kalingrad* or *Baïkonur* could program them into a rapid orbital decay. Low-pass to begin with, once deflected from present orbit, they would enter the earth's gravitational field within three to four orbits and disintegrate upon contact with the atmosphere."

"How long would that take?"

The military officer looked at the ceiling. "Four orbits maximum, six or seven hours minimum."

"Too long Field Marshall. What else is available?"

Once again the Chief of Staff looked to Manarov for support but received nothing in response.

Trishenko tapped the tabletop impatiently.

"Both pass over *Tyuratam*, Mr. President. They could be hit with laser fire on the next orbit."

"Is there a third option?"

The Field Marshall stared down at the baize surface, his thick fingers tracing a tight pattern on the felt material. When he looked up and pinned Manarov in his gaze, Trishenko felt a sense of deja vu. He had seen the silent communication of conspiracy once too often in that past twelve hours. *We may be the ruling troika and I am at the top of the triangle—but two people share something that I do not . . .*

"Nyet!" was the gruff answer. "You have heard the only options."

As if to change the subject, Korovkin began riffling through a thin leather pouch and brought forth a sheath of papers. "The Americans are quite correct. Each name provided checks out. There is a profile for each party including the woman. All were members of the SPETSNAZ and it seems they are all from the Ukraine. They were together at *Kirovograd* and graduated with honors from the advanced base, *Zheltyye Vody*, receiving training as diversionary troops. All served in Afghanistan for one or more tours and are Ukrainian in origin." The Field Marshall smiled, "However, Mr. President, General Manarov, each name checked also appears on the roster of those listed as Missing In Action and presumed dead."

Disturbed by the similarity of the profiles, Trishenko pointed at the papers. "They seem to have much in common—what else do they have as blood brothers—and sisters?" An avid reader, especially of Russian history and, himself, a native of the Ukrainian steppes, the

199

Russian leader felt a vague feeling of uneasiness growing in the back of his mind.

Korovkin pursed his lips, his neck seemed to swell but he read from the last sheet. "It appears they were all assigned to the *Ikhotniki* brigade at the same time."

"*Ikhotniki,* 'hunters' General Korovkin?"

"*Huntsmen*, Mr. President—an elite unit with strong Ukrainian ties and whose roots go back to the early days of the Great Patriotic War. Its battle streamers are many, as are the number of its men and women who are Heroes of the, ah Soviet Union . . . ."

"What do we know of this man Leonid Arbatov?" queried Trishenko.

Korovkin glanced at the report, thick fingers tracing the copy. "Arbatov appears to have distinguished himself in Angola and Afghanistan. Between tours of duty in those combat zones he was assigned to the Reconnaissance Faculty of the Kiev Higher Combined Arms School and the Special Faculty of the Ryazan Higher Airborne School."

A pattern established, the Russian President anticipated the answer to his next question. "And this man Solomenstev—is it safe to assume he somehow traveled in Arbatov's sweeping orbit?" As an after thought, he asked if there was a picture of Arbatov. Responding in the affirmative, Manarov handed him a packet containing an official I.D., and several candid shots including one taken in Afghanistan.

Flipping to another file, Korovkin looked up. "The sergeant specialist served under Arbatov on both fronts and was posted to Ryazan at the same time as the Colonel."

Unable to able to pin down that which bothered him, Trishenko continued; "And what of the disks Korovkin? Do you have comments · on those devices?" Awaiting a response, he shuffled through the photos, stopping at one providing a close-up. He was struck by the intensity of the man's gaze, square face and military bearing.

Subdued, Field Marshall Korovkin shook his head. "Nyet! They are clever but nothing that has ever been employed by either the GRU or the army. Perhaps General Manarov's smart fellows in the basement at *2 Dzerzhinsky_*Square have something to say about the electronic tags. Perhaps that is how he keeps tabs on his 600,000 operatives world wide."

The sarcasm was barely contained by the Field Marshall's steely voice, his baleful glare pinning the KGB chief in his chair.

Turning to face Russia's ranking INTEL officer, Yuri Trishenko rested his hands on his chest, fingers bridged. "Does your operation use such devices?"

Manarov's forehead glistened. Buying time, he smashed his cigarette into the already filled ashtray. "Da! We have developed such a unit and it has been used on a number of occasions. When we have reason to believe the opposition might substitute their agent for one of ours, a subcutaneous implant has served to provide true identification. Battery operated and needing to be placed by a minor surgical operation, such a device is useful only on a limited scale."

"General Manarov, look at the photograph provided by the Americans—it shows a disk whole and one carefully taken apart. Is this similar to—identical to—or markedly different from the units created for KGB use?" The Russian President picked up the facsimile, reached over and placed it directly in front of Manarov.

The KGB officer started to grasp the photo, thought better of it and left it on the green tabletop. His voice—one of resignation. "It is identical to units produced for the KGB."

*　　　*　　　*

Quinn thought about the miner who crawled from the Alaska Shaft to the bottom of the mine to win a hundred dollar bet. He wondered if he himself would attempt it, drunk or sober, for any amount of money.

Nick, white sheet off and tucked under his woolen Mackinaw, edged forward, light from his helmet bouncing off ice crystals on the ceiling. Still chilled by the surface cold, Nick informed him the temperature would not achieve the uniform 53 degrees until the second level. Behind, Running Fox and young Saatela picked their way over the boulders. Light from Huhta's lamp swung around catching Quinn in the soft glow. He worked his way toward the old man. Ten feet away, he saw Nick smiling, his finger to his lips. Five feet away, Nick pointed to a black hole in the ceiling of the tunnel.

"Ore drop, by gully. Gravity was used whenever we could. In the old days, ore was taken from the drift above, put in dump carts and dropped to the lower level. From there it was moved to the cages, loaded and raised to the Crusher Building behind the Head Frame."

Wally and Running Fox came into the circle of light.

"Straight up and we're at Level One. Goes down this stope twenty paces and you're next step is twenty-three hundred feet long."

Running Fox shook his head, a trace of a smile in his eyes. "*A negotio perambulante is tenebris, libera nos, Domine.*"

The men turned to the Native American.

"Latin—something about walking . . . ." Quinn looked at Jimmie.

"Lord deliver us from the Thing that walketh about in the darkness."

Quinn smiled, Nick and Wally shook their heads.

Nick pointed at Wally and Running Fox. "Keep your voices down. They carry in these tunnels. Start looking for timbers or boxes. You'll find almost anything down here. We needs something to crawl up to the next level."

"There's a ladder back there," said Running Fox, "it's in a side tunnel. Wally, let's go!" Saatela and the Ojibway worked their way back down the stope.

Quinn peered up into the almost vertical shaft.

"That miner who crawled to the bottom must have been something to look at when he hit bottom."

"Jahah! A bloody mess!"

"You looking forward to the ride down Nick?"

"Hell No! I told you, that ride's a real pants pee'er!"

Quinn turned to the miner. "You know, Nick, Running Fox and I could probably do as much damage down there as the four of us—you sure it wouldn't be smart money for you and Wally to stay topside?"

The old Finn brought the *Suomi* up and glared at Quinn, head cocked for a long moment. "You don't think this old Finn Grandpa and his grandkid can pack their own weight down there, eh, Mister Marine?"

Caught off guard, Quinn shook his head. "No Nick, I didn't mean that. You've paid your dues. No man should have to fight—what did you call it—*Talvisota*—more than once."

"No mister, I'm go with you. And young Wally and his injun friend back there, by gully, they're gonna be needed too!"

Alone in his office, Yuri Trishenko removed his glasses, rubbing the deep imprints made by the bridge pads. The incoming reports, although abbreviated by the head of each desk, required hours of intense review. Replacing the glasses, he pulled the file forward requested from the archives. It was the wrap-up report on a whirlwind trip to Washington, Minneapolis and San Francisco made by his predecessor. A professor of economics at Moscow University, he had been invited to make the trip with his wife. Flipping the folio open, he smiled at the American headlines extolling the mission. He moved quickly to the section devoted to the first visit by a Russian leader to the State of Minnesota. He recalled the hectic pace of the hastily called trip but the details came back with the aid of newspaper articles and photos, official and unofficial. Overcast most of the way from Washington, he recalled his impressions of the Wisconsin, then Minnesota land scape as the silver _Illyushin_, on a long final, descended beneath the clouds. Early spring, the fields were just beginning to turn green with the rains. He remembered his wife, tapping his forearm, inviting him to look at the American "_Volga_"—the Mississippi River bifurcating the nation down its middle, directly beneath them. He thumbed through a package of material given to him by the Governor of Minnesota and found what he recalled seeing as his plane left that State, heading for California. He unfolded the map and studied it in detail, finding for himself, the location of Tower-Soudan. About to put the package away, he saw the brown folder and by it's size, knew it contained candid shots taken aboard the aircraft at various stages of the flight. Unwinding the red string, he pulled the black and white photos out and reviewed them quickly, remembering some of the incidents and people. About to replace the packet, his gaze fell upon the last picture. Three men seated in the rear of the aircraft, KGB or military agents in civilian dress, he recalled. Two, smiling, lifted their glasses in a toast to the cameraman. The third, not smiling, apparently not drinking, was one seat behind the two men. _The gaze . . . that look_ . . . a sinking feeling gripped Trishenko. He knew the third man was Arbatov and with that instant bit of knowledge, he knew a Russian plane, under the mantle of diplomatic courtesy, had delivered the supposedly dead veteran of Afghanistan directly into the heartland of America.

203

                    *   *   *

Spread open like carefully carved sections of a grapefruit, the world
occupied three screens at the end of the Situation Room. Stewards had
cleared the garbage. One of Hart's secondary assignments was to make
sure nothing but cups saucers and ashtrays went out the door. A
shredder in the far corner was kept busy destroying papers not deemed
essential to the effort but too sensitive to find life outside the
conference room.

    Air Force Chief of Staff, General Benedetto used the light pointer
to describe what had happened on the ground at Tower-Soudan and
what was going to happen next. "Laser-guided bomb took out the fuel
trucks completely. BRIGHTSTAR confirms the fires are all but out.
Infra-red detectors now have ten tanks showing cold, two still running
hot, including the GSR armor in the shadow of the Head Frame. We
suspect it stopped at the fuel depot, filled up and moved out before the
strike."

    "Phil, what about Guardsmen wounded when the stand down
order was given by Holmes?" The President glanced at his
notes. "There must be dozens still out there."

    "Mixed news, Mr. President. Guardsmen went back in teams of
two as soon as the GSR track moved up the hill and out of range. Two
dozen dead and wounded soldiers have been located and taken to the
high school gymnasium. When choppers can get in, the wounded will
be airlifted out." Benedetto looked away from the screen
momentarily. "The Russians got a little heavy handed with their supply
of ammunition. They blew the top off the high school with
concentrated cannon fire. No additional casualties but, as they say in
snow country, no school on Monday."

    Benedetto paused, glancing at General Henderson. "My
congratulations to you, Jocko. Your friend Quinn has done one hell of a
job cutting the armored defense into ribbons. The people left in the
tanks must be thinking interesting thoughts as cold and isolation take
hold."

    Henderson accepted the compliment on Quinn's behalf with a
gracious nod of his head.

    Learning toward the screen, the President tapped the table. "Let's
talk about the weather. What's the prognosis?"

                            204

"The Air Force has the responsibility for weather forecasting." Benedetto nodded to Yoder who placed an overlay on the projector. The black and white chart was a jumble of barometric lines, tightly spaced, fronts following high-pressure areas and assorted wind-arrows. "There's a possibility of good news!" Benedetto focused the star on Tower-Soudan. "The heart of the storm is almost directly on top of the target. Temperature is five degrees above zero with winds reaching gusts of forty miles per hour in heavy snow—classic blizzard conditions."

"Sweet Jesus!" mumbled the President. "What does that do to the wind chill factor—how does it affect men and machines?"

Benedetto's eyes closed momentarily, when they blinked open, he answered; "At that combination of temperature and winds, wind-chill is about sixty degrees below zero. Heat dissipation is rapid—extremely rapid! Anyone in those tanks without fuel is probably feeling mighty uncomfortable." He shifted the star west of Tower-Soudan. "When the storm cell passes, the Alberta Clipper, as it's called, will bring below zero temperatures to the area. The net effect will put tremendous pressure on anyone out there, friend or foe!"

Taking advantage of his time in the limelight, Benedetto pitched air power again. "Weather notwithstanding, the option of taking out the remaining tanks and APC's, one-by-one, is still viable. What we did to the fuel depot, we can do to the armor that still poses a threat—we have a squadron of Warthogs now on deck in St. Paul!"

The President's response was immediate. "I can't read Quinn's mind but the idea of icing those units down seems to make more sense. We need to get our hands on live prisoners, not barbecued bodies. My designated field commander prevails!"

Yoder spoke up from his position in front of the phone bank.

"Taking out Russian satellites, Phil, we're running short on time. What's your plan?"

Benedetto reached for another overlay, placing it on the flat surface. "Air Force Command at Cheyenne Mountain has the whip on this one." He tapped the dimensional drawing showing the curve of the earth for the Northern Hemisphere. Tracked lines interdicting one another over Minnesota could be seen. "Cosmos 948 and 1222—their trajectories have been well plotted and attack parameters defined.

Data's been fed into the computers of the F15's located now at the SAC base in Minot, North Dakota and Duluth. Computer's in the ASAT's themselves are locked into those on board the aircraft and patched back under the mountain. Once released, Cheyenne will be responsible for last second vectors to the targets."

Tapping his gold pencil, the President called for Benedetto's attention.

"As I recall, Phil, aircraft-delivered anti-satellite missiles haven't been authorized by congress. I realize we have such weapons but what tells you that these strikes are going to do the job? And now we're about to take on *two* satellites, both Russian?"

Benedetto tapped the pointer in the palm of his hands. He glanced at Yoder, who covered his chin, eyes avoiding those of the General. Unable to enlist the moral support of the CIA chief who had lobbied for the program, Benedetto faced the Commander-in-Chief. "We have continued the program within the constraints laid down by your office and congress—flight profiles are continually flown, but actual ASAT launch is strictly a function of computers."

"Video games!" snorted the Commander-in-Chief. Leaning back, the President pursed his lips, his voice tired and resigned. "I signed off on that issue to pacify congress and keep disarmament talks on track. We live with computer games!"

"Mr. President, might I refresh your memory—the Tomahawks that did so well in Iraq—they too were products of some well thought out computer games . . . ." Benedetto's face was impassive.

"Touche` Phil!" The President saluted the Air Force General with the tip of his pen.

One of the half-dozen gray phones suddenly rang. Yoder picked it up, listened and looked across toward the middle of the table. "State Department Mr. President. They have an urgent call from the Israeli Ambassador. He apologizes for the early morning intrusion but he'd like to meet privately with you—as soon as possible—says he can be at the White House in twenty minutes."

Rubbing his chin and feeling the stubble, the President realized he'd gone twenty-four hours with less than an hour's sleep. Automatically, he looked toward Hart, like himself, slightly disheveled and bleary eyed. "What can Aaron want at this hour?"

Hart shook his head and pointed toward the ceiling. "Same thing the press wants. Information! All hell's breaking loose! The atmosphere in Washington is charged and the vibrations are being felt all around the world. There have been too many sudden troop movements, aircraft leaving airbases all over the U.S. It doesn't take much to figure out that the Flash Point is somewhere in Minnesota. The Twin Cities airport, they tell me, looks like a full-blown military base already. And, word is slipping down from up North that something heavy is happening at the mine site. Citizens have been evacuated but they've seen the wounded Guardsmen being taken into the high school. The phone ripple from the area has probably reached Seventh Wave proportions!" Hart glanced at his watch, "In minutes, I expect to hear the Speaker of the House, Senate Majority Leader and who knows who else—will be parked in the Oval Office demanding to know what's going on. That's what Aaron Vadiman wants—inside information. Tell'em to meet with the Deputy Secretary of State or something."

Digital finger slicing across his throat, the President signaled Yoder to refuse the call. He felt a light tap on his right forearm. The Secretary of State handed him a folded piece of yellow foolscap, a flowing "K" inked on the outside. He took the note and read it. *Aaron Vadiman is a good and honorable man. I realize time is dear but please, as a favor to me, talk to him when you can.* It was signed simply, "M. Chen". Re-folding the paper, the President looked down the table at the white-haired professor who watched over the top of his glasses. The President nodded. The old man smiled his thanks.

He looked across the table at the Chairman of the Joint Chiefs of Staff. He was silent for a moment and then, speaking softly asked the question he wasn't sure he wanted to hear answered. "Mr. Chairman, how standeth the nation at arms?"

The Admiral's frown faded at his Commander-in-Chief's choice of words.

"The Union standeth tall, Mr. President." He rose, motioning for Yoder to bring back up the world maps and lower the room lights. "Give me Condition One, Tony."

As Yoder pressed a button all three screens lit up with peacock-blue dots.

"There are the critical areas from a purely military standpoint. The dots represent our presence in the region and the fact that all units in these sectors are ready to receive and respond to a shift to Nuclear Alert status." Montgomery moved to one side so all could see the screens. "Give me Condition Two."

Another button was pressed and instantly, blue dots were joined almost one-for-one by carnelian-red blips. "Red for Russian. It is the determination of all of our land, sea and air commanders that the Russians have matched each escalation of ours with one of their own. Granted, we long since lost track of *who* moved first on a case-by-case basis. The bald fact of the matter is that we've both arrived at full military preparedness just one step shy of Nuclear Alert."

"And, Admiral, just how do you see the next few hours?" The President brought his pen to the ready.

"Threshold of war, Sir. The situation is ripe for chain reaction escalation. Whether we take out the satellites or activate the GIUK minefield, Russia is sure to take at least a matching step. It's more than likely that she'll throw the gauntlet down and we'll be looking at World War III before lunchtime . . . ."

"Mr. President?" Yoder leaned forward, head cocked in the President's direction. "It's time to leave for Andrews."

A sudden tightening of the gut was the President's immediate response to the words he had been anticipating. He looked around the room, face grim. "The Russian President noted the Doomsday plane was enroute to Andrews. He hoped it was coming solely for training purposes." A sudden thought occurred to the President. "Do we know where the Russian President is at this moment?" He directed his question to Yoder.

The CIA chief nodded. "We believe he is still operating out of the Green Room. We have unconfirmed information the Blue Train has left the Kremlin and key administrative personnel are being disbursed to hardened posts outside Moscow."

The idea was forming quickly in the President's mind. He glanced at the time zone clocks and then stood up. For a fraction of a second, he momentarily closed his eyes.

When he spoke, his voice had an iron edge. The words came rapid fire. "Yoder! Format a message to Trishenko for immediate transmission! Tell him I intend to remain in the White House. If he's the man I think I know, he'll stay in the Kremlin." He turned to his Secretary of State. "Sam, contact the Commonwealth Ambassador! Immediately! Tell him to have a military attaché, deputy secretary, himself or his wife ready for pick-up in ten minutes. Whomever he designates will be taken to Andrews and allowed to board the Doomsday 747, which will be flown *back* to Wright-Patterson—sans the President of the United States. My good faith offer to the Commonwealth leaders."

A low murmur overrode the white sound. The President looked right and left. He slammed his fist on the conference table. "Now Yoder! Weller! We'll all burn in hell if we don't try something!" He turned to Hart, eyes flashing. "Call a meeting of the congressional leaders for eleven o'clock and a televised press conference at noon! We're going to admit that one of our facilities is under attack by a terrorist organization. No reference will be made to the SPETSNAZ connection." His head jerked back to Yoder.

"Tony, make sure the same information gets to Trishenko!"

Weller, Yoder and Hart converged on the phone bank.

Turning, a trace of a smile on his face, he motioned to Doctor Chen. "Join me in the annex. We'll call Aaron together."

*   *   *

Reaching down, Quinn pulled Wally up the vertical shaft now turned into a man rise. Headlamps off, using a single flashlight, the four men edged toward the iron cage. Muffled voices could be heard beyond the shaft. Cold damp leached into Quinn's hunting jacket. He shifted the M3 from hand-to-hand, its icy metal penetrating through the thermal glove.

Suddenly, a heavy roar filled the tunnel. Quinn froze in his tracks, just short of the cage. Thirty seconds later, a second blast sent vibrations off the rock walls. "APER's! They must be trying to blow Holmes and his men out of the rocks!"

Motioning for Nick to move quickly, he pointed upwards. Grabbing Wally, he put the younger man behind his grandfather, Running Fox next, taking the last place in line.

Slowly, Nick reached out over the deep, black shaft. A dank smell, unlike that in the tunnel, wafted up from below. Gripping the wet, rust-scaled iron rung, the old Finn pulled himself over and began the climb, the cage now directly at his back—two feet away. Wally hesitated but seeing his grandfather disappear into the void gave him the courage to span the gap. In seconds he too, was on his way up. Running Fox, no strap on his carbine, had tied the rifle in the white sheet. When Saatela's feet reached the level of his eyes, Running Fox leaned out over the black shaft, grasping the safety ladder hammered into the rock wall.

Quinn, anxious to get Holmes on the radio, stared down into the ink darkness, then reached across to purchase a hold on the rungs. Voices on the other side of the cage grew louder.

"Jesus! Look out!" Quinn heard Running Fox hiss the words. Head jerking up, he took the blow on his forehead.

Wally's .357 magnum, slipping from his holster, bounced once, leaving a small gash on Quinn's forehead—his foot slipped off the rung. Instinctively, he hooked his arm between iron and rock. Dropping, the full weight of his body transferred to his left shoulder and a sharp jab of pain caused him to gasp for breath. He hung on, feet groping for a rung to support his weight. A wave of nausea racked his body. The voices were animated. He thought he heard a smattering of Russian. Looking up, vision blurred by the blow, he saw Running Fox pressed to the rungs, trying to merge with the rock wall.

Far down the shaft, the magnum hit the wall once, bounced to the opposite side then, a second later, a single shot filled the air, echoing down the half-mile-long rock pipe. Everyone froze. Nick already on top of the cage, Wally and Running Fox glued to the iron ladder and Quinn, pain shooting through his arm, trying to maintain his grip on the wet iron.

Quinn felt a firm hand on his shoulder. Looking up, he saw the Ojibway. With Jimmie's help, he continued up. The old man and grandson, already on top of the cage, reached out, pulling the two men onto the roof. Thimble Wheel and cable, coated with heavy grease, took up most of the space. A third cannon round above their heads shook the cage. Quinn flinched but was grateful the cannon fire and vibrations masked their own movement. He took the radio from his hunting jacket. The pain in his arm had subsided from sharp to throbbing.

Looking up, snow flakes filtered down through the massive iron girders above. Beyond the iron, gray light of morning and more snow.

"KHE SAHN—GATEKEEPER here! He kept his voice low. "Do you read?"

He double-checked the channel to make sure he was ˏcommunicating on slit trench numbers.

"KHE SAHN here." Reception, coming through a field of iron girders, was scratchy.

"The last round got one of my men."

"Cease fire and fall back over the hill. Let them go. We're in position!"

"Falling back now!"

"BRIGHTSTAR to GATEKEEPER, Urgent! We need to communicate!"

Nick leaned toward Quinn, pointing downward.

"BRIGHTSTAR, we've got company!"

Speaking Russian, the "*Company*" was outside the transportation cage and just a few feet below them. Huddled on the roof, the four waited in the shadowed darkness, cold working into their bones despite the insulated clothing they wore. Quinn heard someone's teeth begin to chatter. He turned, looking over his shoulder from his prone position. Wally, eyes bulging, looked like he was ready to vomit. At the same moment, his grandfather, a veteran of two earlier descents on the top, put his arm around his grandson, pulling him as far from the thimble wheel as possible. When he had Wally under his protective wing, he pulled Running Fox close with the other arm. The Ojibway did not resist. Quinn smiled in the darkness.

They heard the shuffle of feet below, felt the car vibrate as weight was taken on and heard the iron doors slide shut.

The conversation below, in Russian, seemed almost jovial as two men talked. Quinn wondered if Karin was in the cage with them. Or was she already dead? The sharp ring of a phone inside the cage reberverated upwards.

"Da!" was the sole response.

Nick poked Quinn with his boot. He look over at the old Finn, sharing his courage with his grandson and the Indian as best he could. His upward glance told him descent was close at hand. A loud, strident buzzer rang three times in close succession. Suddenly, the cable—inches from their heads, snapped. The cage jerked upwards

211

momentarily and then, with sickening acceleration, headed down. Gripping an angle iron, Quinn felt he was floating skyward as the coffin-shaped cage seemed to dropped out from under his body. The roller coaster in the pit of stomach took over. He stifled the urge to cry out in sheer panic as rock walls slipped by, the damp, black airstream washing over his body—light from above fading quickly. No longer aware of the cold, Quinn closed his eyes and fought the urge to let his bladder go.

*   *   *

Rabbi Rachman turned away from the almost hypnotic desert scene to face Ben-Yaacov returning from the communications room.

"Our preliminary fears are being confirmed, Rabbi. We've intercepted orders going out to reserve forces in Syria and Jordan to report for duty. A 're-call' to test response time."

Lowering himself into the low-slung chair, the Rabbi looked at the Prime Minister. "I suppose that orders will now go out to our reservists? Newton's law again?"

The Prime Minister nodded wearily, his face drawn by the events of the past few hours. "Orders to all reservists, regardless of status, are being drawn up for issuance within the hour. And, the decision was just made, that our inventory of missiles will be dispersed from Dimona and targeted for Damascus, Baghdad, Amman and Cairo. Ambassador Vadiman has been instructed to contact the President to ask how much military assistance can be made available to Israel, However—given the circumstances—no one is sanguine about our prospects. In addition, he's been authorized to offer whatever help we can provide—a sign of good faith on our part but an admittedly hollow one."

The Rabbi's gaze shifted to the spymaster; "Eli Ben-Yaacov, is there any hope at all?"

For several moments the only sound in the entombed bunker was the rustle of the onionskin paper in the artificial breeze. "Only, Rabbi, if the Americans and Russians agree to stop in their tracks, if someone backs down or a third party hands them a face-saving solution."

"Might not Israel be that third party?" The Rabbi's words, so quietly stated, took the Prime Minister and Ben-Yaacov by surprise.

White eyebrows flickering, the Prime Minister asked, "And what chance does a dove of Hebrew origin have of being heard over the cry of the hawk?"

"If Armageddon is but hours away, why *not* send doves bearing olive branches? For the love of all mankind why *not* a dove before a shrieking missile?" The Rabbi stood up and took off his black suit coat, folding it carefully and placing it on the paper-covered table. Neither Prime Minister nor spymaster could recall ever seeing the scholar without the black suit coat.

Ben-Yaacov stared at the man's bony wrists and knew that on one arm was a tattooed serial number. Adding to their surprise, Rachman began slipping the knot of the black cotton tie. Loosened, he pulled it off and folding it twice, laid it on the suit coat. Unbuttoning threadbare shirt cuffs, he made no move to roll them up. Reaching across the coffee table he searched for the earlier documents and facsimile photos. Ben-Yaacov looked at the selection, mystified by Rachman's conduct. Studying the papers for a moment, the Rabbi gave a short, almost convulsive shake of his head.

Ben-Yaacov thought he saw the man's shoulders come together as if removing the suit coat had caused a sudden chill. Rachman placed the Hot Line text back on the table, face down and glanced at the two men.

"Both sides prepare for battle and you, Eli Ben-Yaacov can read the minute-by-minute progression to what might well be the end of the world. What is a man to do?" Rachman's eyes looked to the ceiling, hands turned palm-up.

"Rabbi Rachman, every Prime Minister has asked that an eminent rabbinical scholar be called in when Israel prepares for armed conflict. Eli Ben-Yaacov chose you. He knows of your great love for this country—mankind as well. In moments, we both must return to the Command Center and prepare for all-out war. What are men to do?"

"There could very well be a price to pay." Rachman looked directly at Ben-Yaacov.

Baleboss, eyes burning with curiosity at the enigmatic statement, leaned forward, elbows on his arms. "Are we talking gold or blood. What currency?"

"Secrets, Eli Ben-Yaacov. You may have to sacrifice one of your most closely guarded secrets."

"Our nuclear weapons? Ha! That we have them is no secret to the Americans, Russian or Syrians."

Shaking his head, the Rabbi held up his hand. "No, Eli Ben-Yaacov, I'm talking about your ability to listen to the so-called *Hot Line*." He pointed at the intercepted transmissions.

The Prime Minister looked up at the Rabbi, then quickly turned to the spymaster, a blank look on his surprised face.

"From the look on your faces, Mr. Prime Minister, General Yaacov, I gather the coin is of great value?"

"That coin—as you call it," sputtered the Prime Minister, "has purchased salvation for Israel in the past. Asking that it be cashed in is asking much!"

"Ah, but Eli Ben-Yaacov, our spymaster who stands on the same ground as *Joshua*, would you not be willing to give that coin to save not only Israel but perhaps the world as well? If you save your people today, who is to say you cannot re-create the coin tomorrow—perhaps one with even greater power?" The Rabbi reached over and patted Ben-Yaacov's hand in a fatherly manner. "Besides, what is the secret worth the day after the holocaust of holocausts—could it buy a scrap of unleavened bread untainted by deadly chemicals?"

Ben-Yaacov gripped the Rabbi's hand in his. "Placing me in the company of *Joshua* is an undeserved honor but you are right—a coin only has value if used in the right place and the right time. Kept tightly gripped in a sweaty palm—or a wine jar deep underground—it is of no earthly use. What is it you propose?"

"There is a good possibility, *I*, Rabbi Rachman, know why the Russians are vague about their involvement in this matter of American secrets. If I am correct, word must be gotten out to both sides. And once that is done, your secret will be exposed. How else could you be privy to such closely held information?" He tapped the intercepts with a bony finger.

Ben-Yaacov nodded, face somber.

He flipped the documents over, facing the two men. "That time is measured in minutes, I fully understand."

Then he unbuttoned his shirt and removed it, revealing a sleeveless cotton undershirt over a sunken chest. The tattoo was visible. Pulling up the thin cotton garment to his shoulders, he turned, revealing his back—the mark of the lash identical in pattern to the photo laying on the coffee table. Turning slowly from the desert scene, letting his cotton underwear drop over his thin torso, Rachman's voice was strained. "Now I will tell you about the *Pokutnyky* of the Ukraine."

*　　　　*　　　　*

The whine of the cable and spinning iron thimble wheel, inches above his head was strong incentive to keep his face pressed to the oil covered iron roof of the cage. Needing to see something—to focus on one thing as the cage hurtled down, Quinn turned his head, looking up. The transit down, according to Nick would last two minutes and ten seconds. It seemed an eternity. The stop was like the start; a rapid deceleration with a slight yo-yo motion at the end.

"Move! Move!" the gruff male voice was clear.

"Keep your hands off, Soldier!"

Karin's voice rose up from below.

Quinn's head dropped to his forearm—he said a quiet prayer that she was still alive.

The sound of boots on iron could be heard below, then quickly faded. The four men waited in silence for several minutes. Quinn felt the cage move slightly, then Nick's hand on his shoulder. A dim light from the tunnel just below them cast a shadow over the old man's face. He pointed skyward.

"Back on the ladder, ups we crawl to Level 25!"

Following the lead of the veteran miner, Quinn, arm still throbbing, made his way up the iron ladder—Wally and Running Fox in the space between himself and Nick. He felt something brush the back of his neck and turned to see a bat flutter around his head. Forty feet up the ladder, he felt Nick's hand on his forearm.

"Careful, Colonel, there's a narrow ledge and a hand rail. You gotta'works your way around to the tunnel entrance." Shuffling on the six-inch iron plate, he made his way around two sides of the shaft and was pulled into the entrance of Level 25 by Running Fox and Saatela. Just before entering the tunnel, he looked up the vertical shaft to velvet darkness.

Nick switched on his headlamp. Then the lights from all their lamps gave them a view of Level 25. Broken timbers, empty wooden boxes and sections of iron rail littered the tunnel. Abandoned power cables and hoses snaked through the dark along side the track—detritus of a once thriving iron mine. At Quinn's command, they tucked their makeshift parkas under their coats and squatted on their heels or sat on wooden dynamite boxes. Quinn sat with his back to the rock wall, rubbing his shoulder through his hunting jacket, now stained with black grease. Gathered around him, his men, like him, where drained by the terror of the downward descent and the laborious ascent up the narrow ladder.

"Level 25 Nick, that puts us two levels above the Project. What's your reasoning?"

"Not so bad as you thinks, Colonel. We're at the top of the Project. Down this tunnel about seventy feet, we comes to a short passage that's boarded up. When they had us blast the chamber, we had to works our way in from the last three stopes. We takes down the boards, move forward about forty feet, and we be on top of a three-story building built into the chamber. They needed sleeping and eating places and something they call, 'clean rooms.' If we crawls over the roof, we can look down into the entire room and—we might be able to get down to floor level between the back of the building and the rock wall."

Glancing at his watch, Quinn pushed himself to his feet. "Lead the way, Nick! Everybody else—lamps out!"

Turning off the main tunnel and five minutes down the stope, Quinn could see the thick planking wedged between two square timbers ahead.

Nick shook his head "Twelve-by-two-inch, rough-cut planking. Too damn close to blast Colonel. We gives ourselves away before we do any damage."

Running Fox staggered forward, a large object in his hands, behind him, Saatela was dragging a square shoring timber. Nick moved out to help Running Fox. Minutes later, the scavenged timber was next to the sealed entrance. The huge screwtype shoring jack, resting on the tunnel floor, the timber itself, lay between their feet.

The two young men, working as a team, turned the timber sideways, parallel to the planking. Loose rock the size of basketballs were used to lock the timber firmly in place. Quinn and Nick smiled to one another. In less than five minutes, the two men were turning the jack, on its side, with a length of pipe. A minute later, the iron pad of the jack splintered through the plank sealing off the main chamber. A dim light could be seen at the end of the tunnel.

Quinn parceled out the remaining grenades evenly. Lamps off, Running Fox went through followed by Wally and Quinn. The supply bag was passed in and Nick, small, but bigger in girth, was pulled through by Jimmie and his grandson. Crouching, Nick again took the lead. At the end of the tunnel, the roof of the great chambers, sprayed with shotcrete, reflected lights suspended below. A thick plank rested between the small tunnel and the roof of the building built within the chamber.

Quinn looked across the roof and then worked his way slowly over the narrow plank. Sounds of human activity below floated toward the curved ceiling. At the lip, he placed the M3 at his side.

Blue-gray computers, designated Y-MP2E Super Computers, hexagon-shaped and arranged in block pattern, were ringed by the hostages. He recognized the yellow blobs of plastique plastered to the computers, wired one to the other. The prisoners sat on the floor paired off and tied back-to-back. Between the building and hostages, armed Russians wearing American uniforms, stood.

The Russian Commander was using the hostages as a defense against the computers being damaged. To the far left of the huge room, he saw a green canvas shroud. The shapes beneath spoke of the early reprisals. He spotted Nick's friend but couldn't find the dark-skinned woman. At that moment, he saw one of the prisoners' look up in his direction. At the same moment he saw the guard turn to follow the man's line of sight. Quickly he pulled back from the edge.

Where in hell is Karin? Over the hum, he heard voices. The murmur below started to swell. *No choice! I've gotta fall back.* Moving quickly, he scuttled back across the wooden plank to the tunnel.

Nick, Wally and Jimmie, crouched against the wall looked up expectantly, waiting to hear what was happening inside the chamber.

"Hostages surround the computers—tied together. I saw your friend, Nick, but couldn't spot your sister, Jimmie. A prisoner saw me, then a guard. We're in trouble!" Quinn leaned against the damp wall, his shoulder throbbing from exertion.

* * *

The huge hangar door creaked opened like an accordion. Snow gusted in as heat raced out.

The tower in Duluth gave orders to the plows and cleared CELESTIAL EAGLE for immediate takeoff. From a dead stop, the fighter-bomber kicked in its afterburner. Gray Wolf watched the blip move skyward on radar in a near vertical climb. The slant windows of the tower shook under the vibration of the piloted missile launch.

* * *

"If we drop into the drifts we may never get back into the chamber. If we stay here, we may all end up dead!" Quinn looked at his men.

Nick motioned Quinn and the others to the edge of the tunnel and the narrow, dark space between building and chamber wall. "If we moves fast, we can slip down by pushing our backs against the building using our feet against the rough wall. Somebody will have to stay back and hold off the Russians—or lead them away!" He looked at Quinn, then the others. Wally Saatela raised his hand.

Quinn nodded. "Nothing heroic. Just draw them off. Move around. You got that?" Saatela, pocketing the grenades, nodded.

Nick leaned over and gave his grandson a quick squeeze, mumbling an old Finnish blessing in his ear.

Wally extended his hand to Jimmie. The *niitas* locked thumbs and palms together. Seconds later he was out of sight.

When Quinn turned, Nick had already dropped into the narrow space, a two-legged mole working his way down the vertical shaft. Quinn and the Ojibway went over together. Slowly, backs pressed hard against the smooth metal surface of the building, weapons cradled in their mid-sections, feet against the lumpy concrete surface, they slowly worked their way down, three men on invisible chairs, arms spread out to hold themselves like flies against the wall, dropping into the man-made crevice. Beads of sweat broke out on Quinn's forehead as the pain in his torn shoulder shot through his body. The exertion taxed his body and brain. He forced himself to think about the next move but it was all he could do to slide, grope for a foothold, slide and find another spot to plant his feet. Pausing to recover his strength, he saw Nick's white hair ten feet lower and to his left. Looking right, he was surprised to see Running Fox, a dim shape close to the bottom. Suddenly, there were sounds above his head.

Freezing, he looked up. The plank was directly overhead and he saw two forms cross over, speaking rapidly in Russian. Small rock fragments fell on his face, he blinked. Leg muscles screaming in pain, he wedged himself tighter, trying to make himself invisible and relieve body weight pressure from his legs. When he looked up, no one looked down and he saw nothing of a gun barrel poking in his direction. Wondering if he could make it the last ten feet, he gritted his teeth and dropped another foot, took a deep breath and then another. His buttocks suddenly came to rest on an angled incline.

"We've got you," He felt hands on his legs, under his thighs. He relaxed his leg muscles and allowed himself to be lowered to the rock floor. So cramped he couldn't stand, he collapsed to the cement surface. The plank above moved, followed by another shower of rock fragments. Running Fox pulled Quinn to his feet, and held him against the corrugated metal building. A beam of light filled the narrow gap. Breathing heavily, the three men stood pinned under the metal drip shield. The light beam moved slowly from side-to-side. Quinn looked at the twenty-four inch space at his feet for evidence of their presence. Before he could confirm nothing was there to give them away, a dull explosion filled the air above their heads.

Cursing followed. The light seemed to disappear back into the tunnel. Running Fox whispered in his ear. "A trip line tied to a grenade. That could take some heat off."

Quinn felt Nick tug at his other arm. "It's narrow but big enough to work our way through." He pointed toward the base of the metal structure.

Through a louvered slit, they could see the open area. The hostages, now on their feet, still tied back-to-back, had been pushed up against the computer main frames. Two Russians with guns leveled and eyes alert stood between them and the prisoners. They were less than thirty yards away.

*   *   *

"You are one of them?" The Prime Minister's voice was strained.

Rabbi Rachman, buttoned up his shirt and sat down. He grasped the facsimile picture of the Russians in his hands. "Seeing this earlier, made me physically ill. It brings back memories, all of them painful. But to your question, no, I am not of the *Pokutnyky* persuasion. What I know of them, however, might well be this country's salvation."

Ben-Yaacov pointed at the clocks. "Salvation had best be close at hand, Rabbi. We are rushing men and short-range missiles to interdict points at our borders.

In addition, we're ready to commit all of our pilotless surveillance aircraft to find where enemy forces are being massed." He looked at his watch. "And at this moment a coded hotline intercept is being processed. I must also inform you of this: disclosing that we have penetrated the Hot Line circuit is a violation of the Israeli constitution . . . ."

*   *   *

The President pointed to the gray phone. "Supposedly secure, Doctor Chen, but I'd be guarded in what you say."

Chen's small hand came down on the phone. He tucked it to his ear. "*Shalom*, Mr. Ambassador, it is good to hear your voice." He smiled at the President. "Yes, I will call you Aaron and yes, the President is with me."

Matteo Chen grunted but it was a one-sided conversation. On the last note of the five-minute call, his face brightened.

"Yes, he will talk, yes to the last request as well, but we must rejoin our colleagues in a moment or two." He handed the phone to the President.

Hand rubbing his left knee, the President cradled the phone to his ear.

"Good morning Aaron! I wanted you to hear from me personally. We have your best interest close to our hearts." Lips pursed, the President's head bobbed as the Israeli Ambassador spoke at a rapid clip. "Be assured Aaron. *Shalom*—again." He put the phone in the cradle and looked at the head of the Defense Analysis Institute.

"As we suspected, Mr. President, the Israeli intelligence community has picked up on the electricity in the air. To condense his conversation to a mere fragment, Aaron wants to know how much assistance they can expect and when. In return, he offers whatever

220

military support his government can spare—he also reminds us of the fact Israel held off attacking Iraq after being showered with missiles—he thinks it's time for the favor to be repaid . . ."

Laughing softly, the President patted Chen's knee. "Same response whenever the noose tightens. But if we feel the pressure here, what must it be like in Israel?" He rose to leave.

"His last item of information, Mr. President. Aaron said they may be able to serve as a buffer between our adversary and us. He wants to be able to call me back directly. I said yes. You have no objections?"

"None, Doctor, but come, we must return."

*　　　*　　　*

The President was informed that the deputy secretary of the Russian Embassy, along with his senior military attache, had been delivered to Andrews. The Doomsday Aircraft was airborne. There was no response from Trishenko. The President authorized the destruction of the satellites.

The order from Cheyenne Mountain became effective. The "*Nett Lake ricing cano*e" separated from the Strike Eagle as scheduled.

Canadian Air Force radar operators outside of Winnipeg, given a veiled description of a U.S. Military test taking place over Northern Minnesota and Michigan, were surprised to see the blip separate, one accelerating over their border in an easily defined configuration of an air-launched missile. Seconds later, there was another separation.

Messages from Canadian Air Force bases in Central Canada were forwarded to Ottawa, studied and immediately passed to the Pentagon with a sharply worded demand for an explanation.

*　　　*　　　*

"Mr. President, your decision to remain in the Kremlin while they were ordered to the Blue Train was not well received." Field Marshall Korovkin stared across the table at the Commonwealth leader. "To the other President's, the American President's offer to stay in the White House, to send his 'Doomsday' aircraft with our observers aboard—back to its base—is perceived as nothing more than a ploy to buy time and divert our attention."

"A ploy? Perhaps." Trishenko drummed the green surface with his fingers. "But remember that time is relative. If your ballistic attack people are right, we have thirty minutes to seek safety below the

221

Kremlin. If your numbers are correct the White House can be rubble in fifteen minutes, thanks to our submarines in the mid-Atlantic."

The Russian leader reached for the dark wooden pen retrieved from his private suite during a brief respite for a shower and change of clothes. He rubbed the rosewood exterior. Used by the President during his first year in office, it had been a spontaneous gift given at Camp David.

A buzz preceded the start of the facsimile machine. Rising, he took his position at the unit, slipped on his glasses and began to read the third Hot Line message received within the last ninety minutes.

Message complete, he turned to his military and intelligence officers. "I am surprised my friends. They openly admit to having destroyed our Kosmos 1222 satellite but failed to take out Kosmos 948. On the next orbit they will try again."

"Most unusual," rumbled Korovkin, "most unusual."

Looking up, the Russian President was surprised by the lack of hostility in the Field Marshall's voice. He studied the man's face and a feeling of apprehension began to grow. Absence of angst, the man's air of equanimity disturbed him. *He believes that war is inevitable. It is simply a matter of time and he, not I, will be in charge. Korovkin will be the Supreme Leader, the STAVKA will be operational and I will be dead or—a cipher.*

"Most unusual," continued the Field Marshall, "because with the same amount of effort they could have as easily taken out *OKEAN—*Kosmos 1904. They surely know that device is a very sophisticated satellite—our direct response to their LaCrosse spy-in-the-sky. Instead, they challenge us by blasting a weather satellite—its useful days limited. To compound matters, they are going to repeat the process!"

Pulling the third flimsy from his stack, Trishenko looked over his glasses. "Gentlemen, when you consider that action with this report on the SPETSNAZ, does it not occur to you that there is a missing element? Something is not right with this situation?" Unwilling to disclose the knowledge that Arbatov had been in his company when his entourage visited Minnesota, the Russian leader had already pondered what means he had to learn more about Arbatov without going through the KGB.

Korovkin's huge hand slapped the felt-topped table. "I know this Mr. President—the Americans have a problem or have created a problem—it is immaterial. What is of concern is that they are making it the stalking horse for war. You say the President is just, honorable and a man of peace. Perhaps that is so. But remember that he is one man among many. Surrounding him are military and non-military advisers who have led him to the edge. The facts are plain. Since the '100-Hour War' as they glibly call it, they feel invincible. The Americans want war—and they want it now!" The fist slammed down again. An open bottle of mineral water tipped over. No one made a move to set it right.

Looking up at the clocks, Yuri Trishenko moved his chair closer to the table and removed his glasses. "We will send a response over the line. Before we do, however, I have several questions." He turned to face the military commander. "Field Marshall Korovkin, earlier I asked what means we had to destroy our own satellites. Have I been given all the information regarding these options?"

The Field Marshall's neck was swelling, changing colors from white to a pink blush. "An oversight, Mr. President. Since the episode with our nuclear units dropping onto Canadian soil a number of years ago, all units are equipped with a radio-controlled detonation device which can be used to destroy satellites containing either potentially damaging radioactive materials—or sensitive equipment we would not want to fall into enemy hands. These two satellites obviously do not warrant such devices. An oversight on my part." A smirk flickered across the Field Marshall's face.

"One last question, Sergei." The President faced the Marshall. "The report from the GRU, in response to your request, states that *no* SPETSNAZ mission was authorized by any member of any military organization within the Commonwealth. Do you stand behind that report prepared by Military Intelligence?"

Eyes blazing at the continued assault on his personal integrity—and that of the entire Army—he responded with a thundering "Da!"

Swiveling, the Russian leader fixed Manarov with a piercing stare. Tapping the pale green flimsy, signed by the KGB chief himself, he repeated the question. "And Viktor, do you still vouch for your report—deny any involvement in a SPETSNAZ mission?"

"Da!" Manarov's emphatic reply.

223

"Rabbi Rachman, the Harpies of War circle their cauldrons in frenzied anticipation. But the Prime Minister and I must understand who these people are—be able to explain to the Americans and the Russians their presence on US soil as SPETSNAZ."

Rachman's bent fingers tapped against one another. "I can give you a fair portrayal of *who* they are. *Why* they are engaged in this conduct is beyond my ability to fathom. There must be something in that part of the world that they need or want—something that you, Eli Ben-Yaacov, must be professionally curious about?" The Rabbi's gaze met that of the *baleboss* who nodded, lips pursed.

"So, Rabbi Rachman, if you are not *Pokutnyky*, how do you know about them—who are they—and what can we tell Bear and Bull?" The Prime Minister, elbows on his knees, hands clasped, stared at the scholar.

When he started to speak it was as if he were drawing memory from a deep well. "I was fourteen in 1943. Despite the wartime conditions, I survived the Nazis because we lived in a remote area outside of Vilna, then part of Lithuania. Always on the run, one step ahead of the *Gestapo's* dreaded *Einsatzgruppen B*, we managed to outwit the Germans until our band was finally betrayed by peasants who feared they would die for harboring us any longer. At that time, the Germans vacillated: kill the Jews or put us to work. Healthy, I was kept alive and sent to a series of work camps eventually ending up with other Jews and captured Russian soldiers, at *Sobibor*. It was there I was introduced to the *Pokutnyky*."

"If you survived *Sobibor*, Rabbi Rachman, you truly must have been in the palm of God's hand!" The spymaster looked at the Rabbi in wonderment.

Rachman shook his head sadly, "There were times when I would have willed God to squeeze his hand and end my misery. The Germans had discovered the *Pokutnyky* but never really understood that they were a religious sect. They were amused, however, by the stigmata. And that became the genesis of a game—prisoners were pawns. Selected at random, we were ordered to strip naked and handed leather whips studded with bits of barbed wire. In front of the other prisoners, men and women alike, the chosen ones were ordered to flagellate

224

themselves all the while, chanting, *I killed Christ, I killed Christ*. If a Penitent was not enthusiastic enough, the guards would take over—or order prisoners to whip each other. There were other variations, but the procedure was basically the same—tear the flesh from the prisoner's backs until they fell. Some lived, but most did not. I survived my time on the whipping grounds. A member of the sect cared for me. He was a priest-penitent. Through this man, *Aleksandr Binyaminov*, I learned something of their beliefs, philosophy and historical roots."

"And that faith, that philosophy, is based on what Rabbi?" The Prime Minister stared intently at the scholar.

"Judaic roots which trace back to the Pharisees. In doctrine, inclined toward fatalism but at the same time strongly upholding resurrection and future life. The long and tortured path leading to the Ukraine begins with the Diaspora. *Aleksandr* said his people moved northward along the Euphrates in stages. Unwilling to convert to Islam, they were driven from Turkey, fleeing to Armenia around 200 AD When Armenia converted to Christianity they felt compelled to move again. They referred to themselves as *'Job's People.'* They kept the basic tenets of their faith but constant exposure to other religions slowly had its effect. Eventually they found themselves in the south of Russia. When Grand Prince Vladimir of Kiev forced on his people, the acceptance of Christianity in 1000 AD, *Job's People* split into two groups. One converted, the other fled, re-tracing their path south—back to Armenia. Failing in their attempt to remain in Armenia, the group moved back to Russia. By this time, through long exposure to Islamic culture, they were imbued with a warlike spirit—having had to fight for survival every step of the way. They re-joined their brethren in Russia who were Catholic at least in name. According to *Aleksandr*, the lingering effects of these *Khazars*, who were deeply influenced by Islamic teachings, also had an impact on the converts. When re-unification took place between those who stayed and those who left, two important elements merged; the militaristic spirit of the group that ventured south and the effect the Council of Trent had had on those who remained in Russia."

"These people, are they Catholic, Jew or Muslim?" asked the Prime Minister, head shaking in disbelief.

"Today, a combination of all three," responded the rabbi. "The Council of Trent, taking place over a period of years in the fifteen hundreds seems to be the prime catalyst for their present day beliefs. The sacrament of penance established by Rome was like a message from God straight to Job's People. It provided all they hungered for—absolution of sins through three acts—contrition, confession and satisfaction. They believed that God was just, so their calamities could only be the results of their sins. And, the evil could now be averted by confession and prayer. To this they added something not mandated by the Council—contrition through self-mortification. Reaching back into their Judaic background, they borrowed from the fifth tractate in the Mishnah Talmud's. Like the tractate, ' . . . *no more than forty lashes shall be administered. The exact count shall be thirty-nine to avoid the danger of exceeding forty.*' The Rabbi was silent for a moment.

"And thus you have the key—*Pokutnyky* means 'penitent'."

Prime Minister and *Baleboss* looked at each, heads shaking.

Rabbi Rachman paused as he looked toward the desert scene for a moment. "It is interesting to consider that the word 'Islam,' in Arabic, means submission or obedience."

He turned away and continued. "Their survival skills saw them through the persecutions in the Ukraine in 1648. The next three hundred years inter-marriage between this small sect and the Mountain Jews or *Tats* who reside in *Dagestan* and the East Caucasus took place. With the proclamation of Czar Nicholas I in 1829, creating the 'Pale of Settlement' these people found themselves locked into the Ukraine. Their synagogues resemble mosques according to my friend, and their names are mostly biblical to which the Russians suffix '*ov*' was added. Loyal to Mother Russia, they were eagerly sought as warriors and fought with the Red Army in 1917 with great honor."

Ben-Yaacov rose, a troubled look on his face. "But will the Americans or Russians accept this as proof that a small sect of three-headed religious zealots have caused them to go to each other's throats?" He turned to the Israeli leader. "What do you think Moshe?"

The Prime Minister nodded. "The word is, I think—linkage. What ties them to this present act?"

Rabbi Rachman sat back, hands steepled, fingertips touching his lips. "Ah yes, '*linkage*', a most important word these days." A trace of a smile crossed his face as he continued. "Three years ago, at a symposium on world religions in Lucerne, I stood alone late one afternoon looking out over the lake. I was surprised when a hand touched my shoulder. A man's voice said my name. Even before turning, I knew it was *Binyaminov*, Penitent priest who saved my life and helped me escape during the breakout. We embraced, tears streaming down our faces. We talked for an hour. When I asked about his life, I realized that for some, escape from *Sobibor* was to escape to yet another prison. Accused by the Russians of being a collaborator, he spent twenty years in the Gulag. Always the survivor, he was finally released and allowed to return to Kiev. When I asked if he followed the tenets of his religion he smiled and nodded. I asked him if there was any chance for his religion to grow in a God-less country. Their numbers, he said were increasing. In the years following World War II, survivors like himself, took over the leadership of the group and decreed that the young men and women of the sect would be encouraged to seek careers in the Soviet military in order to acquire skills that could be passed on to other members. Of particular interest was having members serve in Angola, later Afghanistan. Those being groomed for leadership roles were encouraged to seek admission in a branch of the Soviet armed services in which the sect had members in key positions. Once assigned to this particular group, advancement was assured. Ranking officers were Penitents. This, according to *Aleksandr*—was the cadre that would eventually allow them to achieve their *'long sought goals'*. The branch of the service his people served in meant nothing to me then. Today I know better. They are posted to the *Voiska Spetsialnovo Nasnacheniya* or SPETSNAZ. Once through the rigorous training, they are assigned to the *Ikhotniki,* the SPETSNAZ's most elite brigade, according to *Aleksandr* . . . ."

*　　　　*　　　　*

Quinn stared through the narrow slit—mind seeking a solution not offered by the eye. Discouraged, he collapsed against the wall of the building. He reflected on the irony of the situation—so close but still stalemated. Running Fox, who had not moved from his kneeling position, gaze locked on the hostages, said "She lives, my sister lives," his words a bare whisper.

Quinn moved back to the slit. Nick scrambled beside them.

"Do you see her?" asked the old Finn, eyes scanning the hostages.

"She is somewhere else but she knows I am close."

Quinn felt a cold shiver run down his spine. Running Fox zipped his snowmobile suit down, reached for the button of his wool shirt. In the narrow band of light, Quinn could see Jimmie finger an ivory-colored, elongated object that hung around his neck. Turning again, his back now to the slit, facing the metal base of the building, Running Fox concentrated his eyes on the dull surface. "Our great-great-grandfather, Chief Adrian Boshey-Little Bear, wore these teeth of a rogue bear he destroyed single-handedly. They are passed from generation to generation. They link us not only to the past but to each other."

Quinn was desperate. He reached for Huhta. "Nick, when you worked on the chamber, did you have anything to do with the building?" His hand tapped the metal panel.

Nick shook his head. "Only to watch it going ups. We blasted this end of the chamber first and cleared it. Building was put together in pieces. Remember, nothing comes down here any larger that what fits into the cages!"

"Think back. What did it look like as it was going up?"

Eyes closing in concentration, the old Finn rubbed his chin. "Three and a half levels! Bottom boxed and electrical, water pumps and heat units contained in the half-level on the bottom." He tapped the metal wall with his hand. "First full level, just above our heads is first aid, the cafeteria—where Mrs. Pavo works and a rest area. Second level is all business—egg heads and their toys. Third level is like a dormitory—open spaces and some smaller rooms for visiting big wigs."

Quinn and Running Fox smiled. Quinn's eyes closed. He tried to visualize the building's internal structure. Eyes snapping open, he grasped Nick's forearm. "Central core! Does that module on the bottom repeat itself on each level?"

Nick nodded. "Jah! Stacked up. A small room inside of a bigger one. Must be an area about fifteen-by-fifteen feet. It goes all the way up."

Quinn looked at Running Fox who nodded in silent agreement. "Nick, you old son-a-bitch, you've just told us how to get into the building. Quick, spread out along the base. Look for a panel that's loose or that we can get a wedge into!" The three men split apart, crawling on hands and knees along the base of the complex. Hands moving over metallic surface. Quinn felt for gaps along the edges. It was a smooth, snug fit. His temporary ebullience was fading.

Nick and Running Fox came toward the middle. Nick reported no success. Jimmie echoed the old Finn, but pulled out the K-Bar borrowed from Quinn. "It's aluminum, probably sixteen gauge. Let's go through it!"

Quinn patted Jimmie's shoulder.

The Ojibway gripped the knife in both hands. Nick eased Quinn aside and put his hands around Jimmie's. Forces joined, they used their upper torso strength to penetrate the metal.

"Holds it right there." Nick ordered. Quickly, the old man lay on his back, his left foot coming up on the spine of the blade.

"Holds it tight, Jimbo!" he hissed under his breath.

Quinn suppressed a smile at the sight of Huhta and the Indian, working as a team.

"We gotter'going!" Nick snorted, his short, muscular leg flexing out. The K-Bar sliced smoothly through the metal for a distance of two feet. Jimmie adjusted the angle and the two men repeated the maneuver. Two minutes later, the return horizontal cut was completed and the aluminum panel bent back. Two brown bats raced out. Quinn flashed a light inside. Hundreds of bats still hung underneath the service building. Dead ahead, the metal-paneled base of the internal module glistened.

Running Fox slipped through the opening, followed by Nick.

Turning to the chamber area one last time, Quinn peered through the slit. The hostages, all seemed awake and alert. He counted guards. Only two remained. Suddenly a single shot rang out, the sound reberverating off the concrete walls. He saw the hostages flinch. A second shot came quickly. He followed their gaze to his far left. Two prisoners, still tied back-to-back, lay crumpled at the feet of the third guard. Teeth grinding in rage, Quinn pushed the nose of the M3 through the slit at the executioner. Flipping the safety back, his finger came to rest on the trigger. Mentally calculating the range and the swing time back to the other guards, he remembered the plastique and remaining hostages. Slowly, he pulled the barrel back, shook his head and slipped back through the opening.

The shots and crawling men caused bats to drop by the hundreds. Whirring, flapping sounds filled Quinn's ears as he crawled on his belly to avoid contact. The acrid smell of bat guano clogged his nostrils. Shooting pains radiated down his arm and side. Hands reach out as he reached the base module. He was pulled inside the lower unit to the hum of electrical equipment.

Nick's flashlight snapped on. "No sweat, Colonel! Lifted a panel like taking off a storm window. Jimmie's checking things out."

*   *   *

Rabbi Rachman turned away from the mural. He stared at the spymaster at the open door, somber-faced and now in his working military dress. "So it's come to that, eh?"

Ben-Yaacov nodded. A shoulder strap held a holstered Berreta to his chest. "The situation is very difficult. Ambassador Vadiman has not been able to contact Dr. Chen who must be sequestered in the Crisis Management center. As to the Bear, we've had no success in reaching the President. The Prime Minister has asked our *Charge'd'Affairs* in Moscow to use whatever means possible to accomplish that end— nothing! It would seem, Rabbi, we are at the point of no return! If we cannot communicate with the Americans and Russians our best laid plans are for naught!"

Ben-Yaacov's flight suit on, let his hand come to rest on the Israeli-made Berreta strapped over his heart. "INTEL confirms that guidance radar is being activated in Syria and Jordan but not Egypt. A rocket attack is imminent."

"And you, General, what role will you play in this blood bath?"

"I will lead the airstrike against Damascus. Should I be so lucky as to survive. I'll return here if there is a runway left to land on, re-arm and lead a second wave strike to Amman or Cairo."

The Prime Minister, wearing a dark suit walked in. "*Shalom Aleichem*," he said without enthusiasm. "Doves have been sent out, but like those on Noah's ark they have returned. There is no place to roost." He raised his hands in frustration.

The three men stood a moment in awkward silence.

Ben-Yaacov looked at his watch, the clocks on the wall and then to the two men. About to turn toward the door, the Rabbi reached out, his hand touching Ben-Yaacov's forearm. "Wait! Indulge this old rabbi

230

a brief moment! I am no diplomat, politician, warrior or spy but I think there is a way for the doves to find a roost. Will you listen to my proposal?"

*   *   *

The President had to agree that his decision to stay in the White House had compromised the future of the Country. In six minutes he would face the waiting press. If the Russians authorized a missile strike from the mid-Atlantic, Washington was doomed and the US Government in jeopardy. His effort with the Russian Chief of State had collapsed. Rubbing his throbbing knee, he remembered his commitment to meet Major Hughes and his sons at the Christmas party. A wave of sadness engulfed him.

The Chairman of the Joint Chiefs had the floor. "The only steps left short of going full Nuclear is to authorize me to make contact with our submarine fleet. Should our boomers find that Russians subs within the strike envelope appear to be activating for launch, we take them out!"

The President glanced at the clocks. There was no good reason to continue his chosen course. The situation demanded action. Nothing had happened to justify any course except prevention of a first strike attack on the American mainland. He raised his hand to acknowledge the military's position. "Activate TACAMO."

*   *   *

Quinn saw the dark shape come through the narrow hatch-like opening above. In seconds, Running Fox was at his side. "You were right, Colonel. It goes straight up. Nobody on the first level. It's dark. Second level is lit up and you can see into the printout room. Three Russians in there and two prisoners—Karin and my sister, tied back-to-back behind mattresses in a corner. The Computers must be doing their thing. Charts are pouring out of two machines. Others appear to be tape-driven."

"Arbatov—an older guy with white hair—is there?"

"Nope. Three men, two officers and a sergeant. No eagles in sight. I didn't try Level 3—didn't want to risk being detected."

"Karin and your sister—could we get them out?"

Running Fox shook his head. "Fraid not Colonel—they've got necklaces on—and that explains the mattresses."

231

Quinn looked blankly at the Ojibway. "Each have a grenade around their neck. I'd say the arming pins have been partially pulled and wired. I've seen that technique used by the Viet Cong. If one of the victims moves or falls asleep, the pin comes out and . . ."

"Jesus, sweet Jesus," Quinn muttered quietly.

"One more thing, Colonel. A countdown clock on the wall—I'm going to guess it's timing the computer run."

"What did it read?"

"0007 Minutes and winding down when I left."

*   *   *

"There is absolutely nothing to be gained by such a move!" Field Marshall Korovkin's voice bellowed throughout the room.

In contrast, the Russian leader's voice was calm, almost subdued. "The President has made a token gesture. We will respond in kind. Kosmos 948, is in its fourth year. It would shut down within months. We will send the message. Now!" He pointed to Manarov.

Three phones had been added to the table. Manarov picked up the middle one and gave a terse order. Across Moscow Square a Russian general nodded to the wall, put the phone in its cradle and uttered a single word to the transmission officer. "Proceed."

*   *   *

Nick took Quinn's place on the second level. The old Finn, his knowledge of Russian rusty, was assigned listening post duty. Quinn and Running Fox searched frantically for the independent phone line Nick remembered from mining days. The line bundled with other communications and electrical cables, hundreds in all, snaked through the chamber to a relay station located at the bottom of Shaft Eight—the cage-lifting shaft.

Quinn's eyes scanned the small room. *If the line can be found, if it's still operational and if someone answers it we can make contact with the outside world—and no lights would blink on the communications console for Arbatov to see upstairs.* According to Nick, one phone was located in the Superintendent's home just west of the mine, the other on the desk of Tower-Soudan's chief of police. An early version of the Hot Line, it was designed to be used in the event of a disaster in the mine.

Nick said that it was tested twice a year during mining days and had been incorporated into the tourism and university system. He'd not heard anything about it recently . . . where is the damn thing!

Running Fox, K-Bar in one hand, small pen light in the other, cut apart multi-colored bundles. Quinn, flashlight tucked under his arm, studied the hardbound logbook chained to the work desk suspended from the wall. Pages of entries filled the document. He searched for a clue that would identify their only hope of communicating with the outside world.

Suddenly, a light flashed on outside the module behind them. Quickly, both men snapped flashlights off and froze. Quinn motioned Running Fox to move right, crouching down, he moved left. They met at the doors. Looking through the narrow, vertical windows, a man dressed as an American soldier, stood at the main work area in the kitchen. From the supply of bread and meat in front of him they knew he would be in the kitchen for a considerable period of time.

"I could take him out," whispered Jimmie, the back edge of the K-Bar touching the tip of his nose.

"We can't afford to bring down the wrath of Arbatov," hissed Quinn. "Stay here and keep your eye on him. If necessary act. I'll get back to the phone line."

Seconds later, he was at the work desk, his finger running down dated entries. Sweat trickled down the small of his back as he searched for the key to the ancient phone line and communications beyond DANTE'S ORCHARD. The entry dated 10-21-83 caught his eye. "EmerLine—Tag 101-B/W—OK." Quinn scanned the myriad of colored wires. He found two, black and white, but with different tag numbers. About to go back to the book, he pulled a thick power cable out and felt a thin wire taped to its backside. Twisting the cable brought the wire into view, he saw the tag."EL/101". He cut the electrician's tape binding phone line to power cable. Lifting the slanted desk top, he rummaged in the tool compartment. Anxiety growing again, he threw the work papers and prints on the floor. *There's gotta be a handset in here. God in Heaven let there be a handset.* He lowered the desktop in disgust and stood back, his light playing on the area. Then he saw the black device hanging on a nail under the desk. He grabbed the old fashioned serviceman's handset with the rotary dial mechanism built into the head. Four loose wires hung down—two had plug-in heads, two, copper clips. Moving quickly, he circled back to Running Fox.

"Here's the phone. The line is exposed, tagged EL/101. You make the patch, I'll watch our chef." Jimmie grabbed the handset, disappearing into the darkness. Quinn stood up and looked into the room. The Russian was putting sandwiches into paper sacks. He counted six individual bags. A single large bag was already made up. Only a half dozen men remained? He remembered his last meal— Finnish stew—then realized how long it had been. The chili had been ignored and was charred history. He'd been running on adrenaline. Several minutes later, Running Fox tapped his shoulder.

"The phone's hooked up and the line is live. The directory's on the desk. Good luck!" The Ojibwa slipped back to his watch over the kitchen area.

Out of sight above, Nick lay curled against the wall, ears straining to pick up bits and pieces of conversation.

Quinn dialed the four-digit number listed for Emergency. The device turned, like tumblers of a safe. In the park superintendent's home, the ringing echoed throughout the empty house. Thermostat set at 47 degrees, the residents basked in the high 80's of the sunny Bahamas.

On Chief Hietala's desk, its sound was lost in the background of other phones that never ceased ringing. Grim-faced officers and deputies drank coffee and waited for new orders. The entire community of Soudan had been evacuated to Tower. Wounded guardsmen were being treated in the basement of the high school.

Chief Heitala came into his office. The rookie behind the desk, given the responsibility of fending off reporters, pointed at the phone he'd been told would *not* ring. The Chief stared in amazement. The rookie handed it to his boss.

Quinn heard the click, "Who've I got up there?"

There was a cough. "You got Chief Heitala. Who in hell is this?"

"Colonel Quinn, U.S. Marine Corps. I'm down in the mine. I need your assistance and I need it now."

"How do I know you ain't one of them Russian sonsabitches that's killing our boys?"

Quinn sucked in a deep breath. "You got a call between four and five last night from the FBI right? And they gave you some information about me. Have you still got it?"

Heitala spotted the yellow Post-It pad stuck to the four-drawer file next to his desk. He leaned over toward it.

"The name is Caleb, no middle initial, Quinn. Colonel, USMC, serial number 01620509. Current home, Beatrice Lake—presently known to be in the company of one Karin Maki, daughter of Frank Maturi. Also in the company of Wally Saatela, his grandfather, Nick Huhta, and Jimmie Running Fox. You got enough?"

"Jeeesus Colonel, I'm sorry! What can I do for you?"

"Take this number and call it immediately." He gave Henderson's priority number and some brief instructions. A movement above caught his eye, Nick was silently working his way down the metal stairs.

"You said to comes down if anything changed up there." Nick's haggard, blood stained face was showing the rigors of war. "The big cheese must've been on the third level. He's back, and fit to kill!"

"Did you pick up any of their conversation?"

"Not much. If I picks up word or two—even then, it's strange talk. And they're keeping an eye on that clock, I tell you!"

"Strange talk, Nick? Try me."

"They drinks tea but talks about coffee and bread. When it isn't that, it's looking at the clock."

"*Pita*—they talk *about pita*. That's some kinda bread ain't it?"

Quinn nodded, more confused than enlightened.

"Colonel! You there?" The voice of the police chief boomed in his ear.

Priest rabbi, bread, coffee and clocks. What else is there?

"You OK Colonel? You looks kinda sick."

Rubbing his eyes, Nick came back into focus. "I'm fine. You get back to the listening post. Same routine. Come down if something else comes up."

"Colonel! You there?" Heitala's voice roared through the handset. "I got the number. They want you to stay on the line—said they were gonna try some tricks!" Quinn covered the mouthpiece again and leaned against the wall. *The shawl, skullcap and leather straps. I must be falling out the tree myself. Mother of God help me stay awake.*

Good arm keeping him propped against the desk, Quinn's head dropped. Everything seemed to go gray then slowly to black.

"DO WE HAVE GATEKEEPER? This is BRIGHTSTAR. Hang on, points East coming on line."

"Quinn, Henderson here. How you doing?" The Commandant's husky voice was crystal clear.

Quinn snapped out of his stress-induced reverie.

"*Just* doing. Arbatov's got things going pretty much his way. LaMont's people are tied to the CRAYs—and the CRAYs are plastered with plastique—two hostages directly above me have grenades tucked under their chins, pins partially pulled. We might have iced one more Russian but he's killed two hostages in retaliation. All I can tell you about the computer program is that it's running behind schedule. Other than that, the only way to stop our Russian is to keep him from sending the data. The way I read it, it has to be done from your end. If I make a move a lot of people down here are gonna die. Have you taken out their satellites?"

"Cosmos 1222 is down, Russia just informed us they will destroy Cosmos 948—over the US-so we can see it happen."

"Huh?" Quinn shook his head, straightening up.

"You heard it right."

"That's it then. Somebody tells the SPETSNAZ commander that the war games are over. We bury the dead. Somebody else apologizes to the widows and orphans."

"I understand where the alligators are for you, Caleb, but we don't understand the Russian gambit. We'll blow 948 to hell if they don't. Meanwhile, what can be done to prevent Arbatov transmitting out of the mine?"

"We've got to find the missing Command Track! Remember, Holmes said there were two antennas. One mounted, one crated. They've planned this well. Two dishes must mean primary and backup. If we can take both out Arbatov will have to hand-carry the data out of here. If it comes to that, I'm back in the game. Meanwhile, we follow-through with the ice-down plans. Let the Marine brigade and Norwegian troops slip their leash as soon as weather permits and form a body-to-body cordon around this place. Nothing, not even a white weasel under six feet of snow, gets through the net."

"Fire Tankers are loaded and waiting in the hangars at Brainerd—a Helicopter with a bucket is waiting at Ely. They fly in fifty minutes."

"The weather's improving up there?" He was having trouble keeping track of time.

"Maybe God does look out for widows and orphans Caleb. Weather forecasters say it will be clearing two hours ahead of schedule. The temperature, however, is dropping. Ten additional Chinooks reached Camp Ripley early this morning, loaded up, lifted off and are heading north as we speak. They'll land at Grand Rapids, behind the storm and be in the air again when they can touch down at Soudan. Minnesota Highway Department is clearing a straight section of Highway 169 between the two towns. The balance of the Norwegian-Marine unit will land, unload and take off at three-minute intervals. All 800 men will be in position by noon, your time. Four AV8 Harriers from Cherry Point are on the deck at Holman Field in St. Paul. You'll see them on your picket lines at the same time. Holmes has been given a battlefield commission to major and will function as forward air controller and liaison between Guard, Marines and the Norsk troopers . . ."

He heard Running Fox hiss his name from across the room.

"Down Colonel! Get down!"

Pulling the tap clips free, Quinn dropped to the floor under the worktable. Air vents and conduits in the center of the module preventing him from seeing the door. He heard the sounds of the steel door being rattled followed by the smashing of glass. Bent over, fingers on the floor for balance, he worked around the equipment. In the dark, he saw Running Fox, flat against the wall. He motioned for Quinn to stop in his tracks. A hand reached in, fumbled for the knob and disappeared. Several moments passed. Suddenly, the doors parted with a crash. A chair had been used to clear the opening. Quinn edged back behind a space heater. He brought the M3 up to his good shoulder.

A dark form materialized, backlit by the kitchen lights. It was low to the ground, a second later on the ground.

Coming around the heater, Quinn saw Running Fox withdraw the K-Bar from the Russian's back. Quinn and Running Fox dragged the soldier into the module and the doors swung shut.

"He headed this way. Know you didn't want any killing . . .yet."

*   *   *

Aboard the Russian *Mir* Spacecraft, orbiting at an angle to Kosmos 948 and slicing southeast over Baja, California, the commander keyed twelve digits on the console as the satellite approached the interdict point on the CRT. He flipped the toggle switch upwards. Kosmos 948, at its transit point over Resolution Island, disintegrated in a blinding flash. Hundreds of burning streaks appeared seconds later as fragments hit the atmosphere over the Ungava Peninsula. The fishermen of Belcher Island in Hudson Bay saw the momentary spectacle. A single streamer was reported as far south as Moosonee at the southern tip of James Bay. NORAD flashed confirmation. There were no detectable traces forward of its established trajectory.

Personnel aboard Yoder's quickly re-activated SR71 provided both visual and radar confirmation of its demise.

On level four in the blast-hardened lower levels of Kalingrad Space Center, 15 miles northeast of Moscow, Agricultural Specialist Giorgy Golubov wondered why, within the span of two hours, he had suddenly lost one hundred per cent of his ability to determine the advancing frost line. Mother Russia, he thought, was back to rolling dice to determine when to plant its corn, wheat and millet.

*   *   *

"It's your classic tit for tat. The Americans have sent their Doomsday craft west and the Russians voluntarily destroy a satellite. What it all means is beyond me." Eli Ben-Yaacov, still dressed in his pressure suit, stared at the green flimsies.

"One thing it does mean, Eli, is that we put a hold on the Rabbi's suggestion." Ben-Yaacov looked up at the Premier, not sure whether what he heard was a statement of fact or a thinly veiled question. The Premier continued; "We hold until we can determine if this package of gestures really has any substance to it or it's chicken soup with no matzah balls . . ."

Ben-Yaacov motioned toward the entrance to the main command post. "The air on the other side of that door is charged with electricity. A farting fly could activate the system. Every sign we've ever seen—in any other situation that led us to war—is out there. Intelligence reports from the Palestinian sector in Jerusalem indicate an '*unearthly*' quiet. Off shore, our picket ships have been seen an increased number of

Syrian naval vessels steaming out, with ship-to-ship missiles. Suddenly, we have to split our numbers and look in all direction for potential problems. Only Egypt seems to be out of sync with the Arab World."

"Perhaps the Camp David Accords are worth something after all," murmured Rabbi Rachman.

Elbows on his knees, hands clasped nervously, the Prime Minister eyed the Rabbi. "I admire you Rabbi—your source of optimism seems to flow from a deep well. Do you really believe the doves of peace can find somewhere to land despite all you've heard?"

Rachman's face broke into a faint smile. "They did for Noah . . . finally."

*   *   *

An agitated House and Senate leadership waited in the White House Conference Room, the scheduled meeting and press conference on hold.

The President tried to assess the mood in the Situation Room. There was little elation at the death of Cosmos 948. *Russians voluntarily destroy the second satellite but nothing seemed to change.*

Hart entered the room unsmiling. It had been his assignment to announce the conference delays.

"Mr. President, if the congressional leadership doesn't get our scalp, the networks will! Everyone's demanding answers and explanations. The meeting delays are sticking sideways! There isn't a part of the country that hasn't felt something. And the networks are onto our military movements into Minnesota. We've used weather as a cover to its ultimate. All of Minnesota is closed to air traffic except military. They aren't buying that. I expect to hear reports of private planes and helicopters being used to sneak up north to get a scoop. We don't have element of geography going for us like we did in Kuwait to keep the press out of the picture! Now, they're gonna be in the war zone before our troops arrive!"

"Yes and No," answered the Chairman. "Yes, they can be held back if it affects national security and they know what the stakes are—a combination of military dictate and voluntary agreement. If we continue to keep them in the dark much longer, however, somebody is going to violate protocol and slip through the net."

"Mr. President?"

239

The Commander-in-Chief turned to face the Commandant of the Marine Corps.

"Yes general."

"General Smallzreid and I have been rubbing heads at this end of the table. My men and the Norwegians should be able to create a ground seal around Soudan within ninety minutes. It wouldn't be hermetic but tight enough to keep people from getting in harm's way. Smallzreid tells me he has two squadrons of armed observation helicopters on deck in St. Paul. In addition, Nighthawk helicopters have been off-loaded in Minneapolis and can be airborne immediately. A Special Forces team from Fort Pope has men inside all air-traffic control facilities ringing the Twin Cities and Duluth. Between fixed wing and helicopters—with radar and infrared detectors, we could probably keep most airborne media out of the operational theatre. If you do your jawboning upstairs, ask for their cooperation. Maybe we can keep them out of our hair until this matter gets resolved."

The President nodded his approval. "I'll go up and make the pitch to the press immediately!"

"No, Mr. President!" Hart rose from the table, one hand in the air—palm up. "I'd suggest someone else—preferably non-military—go up and talk to the congressional leadership and press at the same time. We've agreed to the *'unknown terrorist'* ploy—if you do the talking and this thing bounces the wrong way, there's no insulation left around the White House. You're Presidency is out there swinging in a cold wind— by its lonesome."

Heads bobbed in agreement around the table. The avuncular George Weller rose from his place. "I've been sitting on my dufus long enough. While I don't have the President's patented charisma I have been known to peddle a little snake oil in the past." The Secretary of State rose, rolled his sleeves down and slipped his coat on. Adjusting his tie, he looked around the room with a smile. "Keep the lid on, at least until I get out of the hands of the lynch mob." Following the lead of the President, the entire group stood up in silent tribute to the sacrificial lamb.

*　　　*　　　*

On his knees, handphone re-connected, Quinn looked at Running Fox and Nick, just down from the second level. "If I read Arbatov right, he's

240

in charge of men who volunteered for a one-way trip. *They're expendable* but the mission is not. Jimmie, you could be looking at one, possibly two Russians coming down the stairs checking after our dead sandwich maker."

The dark-eyed, striped-faced Ojibwa nodded.

"Nick, are you ready to do some sharp-shooting?"

"Shoot high—short bursts . . . and don't hit the computers."

"Good! You take the fieldphone and wait for my signal. Jimmie, you take out the Russians any way that works. And make that sonofabitch upstairs wonder what's happening. We should nail the guards while the computer is still running the program. That's possibly our only bet to peel the hostages from the CRAYs and the plastique! He put his hand on Jimmie's shoulder. "Your sister and Karin are in deep trouble. I know you want in on the second level but you're needed in the kitchen first. Then work your way up! Nick, you've got five minutes to set up the shooting gallery!"

Running Fox faded around the space heaters, Nick scrambled down into the crawl space. Quinn tested the line.

"GATEKEEPER to BRIGHTSTAR, KHE SAHN who's on line up there?"

"KHE SAHN here, Colonel. Snow is diminishing, visibility increasing. Command Track not seen yet."

The welcome sound of Ernie Holmes voice was followed by a voice from the AWAC.

"BRIGHTSTAR in the blue at 27. Skies clearing to the west with a dealer's choice of airborne ordnance stacked above you. Infra-red indicates last of tanks and APC's have turned cold."

"Jocko here, Caleb. Advance Ripley detachment in-bound on final leg and the Harrier are over Buhl, Minnesota in the center of the Range—water tankers off the deck at Brainerd—Warthogs at the ready. Go ahead!"

Quinn smiled. Contact with the outer world was like a breath of fresh air and a source of renewed energy.

"Jocko, the computer run must be just about done. Arbatov seems to be preparing for *some* means of transmit. The black box transmitter appears to be on the third deck. We're trying to save the hostages in the main chamber. The two women hostages with Arbatov must be his

241

personal life insurance policy. Are you sure you've got the right satellites?"

Mumford responded quickly. "There are no candidate birds left in space—Arbatov must have black boxes inside the black boxes. We're overlooking something!"

Mumford's words sparked Quinn's memory. "KHE SAHN, Ernie, when White Top and his men loaded the Command Track did you get a look at any of the radio gear?"

"Just barely. It was a pretty rushed affair."

"Give us a shot. ROCKAWAY, listen to this. Go KHE SAHN!"

Major Holmes quickly described what he'd seen being placed in the Command Track—now missing.

"Oooh, we got'us a problem," responded Mumford at Langley after hearing the description. "That thumbnail view of KHE SAHN's tells me that your man has got himself a Russian R350M radio—with encryption and burst transmission facilities. If he can get loose, he can play a waiting game and send that data to some satellite that isn't even close to transit orbit over Tower-Soudan or—make some kind of land line or microwave transmission. Are there towers anywhere in the area?"

"KHE SAHN here. I'm looking at the tops of three different towers in my field of view—one to the east looks like it has microwave dishes mounted near the top—others are local radio stations and . . . ."

"Go KHE SAHN—make it quick!"

"I see another antenna—right in front of me Colonel! Somehow, the Russians were able to mount it on a jerry-rigged platform near the top of the Head Frame during the night. It's barely visible but I'm sure it's the mini-dish antenna they tucked in the back of the Command Track. I don't have the firepower to reach it from here. Also, the Russians still control the Engine House in between."

"Ernie, is that antenna aligned with the microwave tower?"

"No. It appears to be fixed facing the northwest—angled up about 30 degrees."

"Jocko here, Caleb. We can take out the Head Frame. I'll execute the . . . ."

"NO! No Jocko! Take out the Head Frame and everyone down here dies. There's no way in hell anyone could get out!" The sudden

242

churning started up again in his gut. *We're expendable—Jesus Christ—we're dammed expendable.* He felt his knees go weak.

"Understand Caleb—but we've got to shut down his ability to transmit."

"For God's sake Jocko! Hasn't that been done? How far away are the Harriers?"

A disembodied voice responded. "Ah . . . make that 58 nautical miles. ETA over Soudan two minutes. They're at loiter speed now, waiting for orders."

Quinn closed his eyes. "Vector two to the mine! Work with Holmes to put one eye-ball-to-eye-ball with the antenna unit in the Head Frame. The second provides suppression fire *at*—but not into the Engine House. Tell the pilot responsible for taking out the antenna he's gotta play brain surgeon. If he can take it out with his down-blast great! If not, he fires minimal machine gun bursts to tear it apart." Quinn glanced at his watch. "I gotta'go! Will make contact ASAP!" Without waiting for a response, he disconnected the clips and raised Nick on the radio. "You in position?"

"Jahah! Two guards standing."

"Can you get them without hitting hostages or the computers?"

"There's about a ten-foot spread. Not much but I can do it!"

Quinn checked his watch again. "Nick, make sure the hostages move away from the computers and that you're not exposed to fire from the second level. Sixty seconds Nick, start counting!"

Quinn moved toward the iron stairs leading to the second level. Half way up and he heard the two shots fired, followed by a short burst of automatic fire. He heard pounding of feet on the floor overhead. Muffled screams in his ears. A second long burst of fire, an entire clip, then he heard the sound of a single round. Flipping M3 to "Fire," Quinn lunged toward the second level deck.

*   *   *

The Old Finn was on his knees, peering through the narrow slit. The head of the closest guard in his gun sights. He made a short practice swing to the second, all the while counting seconds. As a number twenty passed his lips, the roar of gunfire above filled the narrow crawl space. He'd come through the space carefully, so few bats had been disturbed. Now, the air behind him was a mass of whirling brown bodies, flapping and screeching. They beat against his head and shoulders as they fought to get out from under the building. Flinching, he looked out to see the guards drop to their knees, looking directly over his head, weapons locked at the shoulder, seeking a target. Nick's mind raced. If he fired now, the hostages, most of whom had fallen to the floor would be in his direct line of fire. He heard guttural commands directly above his head. The two guards, responding to the orders to spread apart. Suddenly, an old memory came flooding back; Russian infantry trapped in a granite quarry and well hidden. A veteran hunter killed the invisible invaders with gunfire deflected off the rocks that provided them cover.

Nick quickly angled his vintage weapon downward and squeezed off a short burst, swinging the weapon to his left and with the familiar "*brrp, brrp,*" repeated the process. The rounds hit the concrete floor directly in front of the guards. Chips sprayed into the air. Flattened by the impact, the bullets ricocheted wildly, some spinning as velocity decreased. One guard fell a deformed round smashing into his skull like a miniature sledgehammer. The second guard rose, aiming his weapon in his direction. Nick ducked as gunfire raked the base of the building. Then he felt a sharp, burning sensation in his left thigh. It took his breath away, his eyes momentarily winced shut. The Russian was slamming a second clip home. Nick bit his lower lip and fired.

In slow motion, the soldier did a complete back flip, and landed on the green tarp covering the dead hostages.

Bats continued to flutter past his head seeking an outlet in the vast chamber.

"MOVE! MOVE!" Nick heard Running Fox's voice directly over his head ordering the hostages away from the computer array. Leg dragging, Nick worked his way toward the edge of the building. Squeezing out of the small entryway, he looked up. Running Fox, back pressed to the wall on the steps leading up the first level, motioned the hostages away from the hexagon-shaped computers. Tied back-to-back, they moved like crabs, stronger dragging weaker, men dragging women. One large woman, tied to a smaller man, simply carried him on her back as she lurched away from the computers, whose soft blue lights indicated that they were still processing data.

Nick broke into the open, racing toward his friend, the cook. Pain forgotten, he put his shoulder under her arm. Her eyes wide at the sight of her rescuer, the day-shift cook grunted as he cut her feet from her partner. Behind them, the remaining hostages sought cover near the base of the building. Bats crowded the air, their high-pitched sounds and flapping wings drowning out the hum of the computers.

*   *   *

Quinn stared through the narrow windows. Karin, her hunting jacket open stood facing him with her eyes closed. Cot-sized mattresses were piled around her shoulder-high. The grenade hung between her breasts. Skin chalk white, blond hair matted, hands tied behind her back, she seemed frozen in space.

Behind her, Arbatov, in the uniform he wore when Quinn first saw him, stood behind a wide printer that had paper feeding out at a rapid clip. Karin was partially concealed but he could see portions of a graphics presentation feeding into a neat stack. Jimmie's sister was on the other side of the Russian, her back to him. Hands tied like Karin, she was barricaded behind striped mattresses. There was no doubt in his mind that she, too, wore an ornament of the same design and configuration as Karin.

Suddenly, the commander turned.

Quinn edged back from the narrow window. He heard the familiar flat voice. "You're friend, *Nurse* Maki, is very good—a professional soldier of the first order. However, I will prevail. He may have the other hostages but I have you and the data."

Quinn paused, then looked into the room. The countdown clock showed four zeros.

Karin stared at the door as if in a trance. The Russian was watching the printout come to a halt.

Karin blinked once and her eyes opened wide at the sight of Quinn staring at her.

He could see tears suddenly roll down her cheeks. She looked down at the grenade and quickly back to Quinn. He nodded. Slowly, she drew her shoulders up.

He brought his finger to his lips. Tears continued to fall.

"Complete. Finished. The run is over!"

At the sound of Arbatov's voice, her eyes rolled upwards. He stared, not comprehending.

She repeated the motion. He shook his head.

Karin's head slowly cocked left, eyes rolling toward the floor. Her head straightened up and the eyes moved skyward. Quinn nodded. He pointed at the floor behind her, toward the graphic printout then toward the ceiling. Eyes focused straight on him she nodded almost imperceptibly.

The black box gear is upstairs.

A sharp, "crack" made the metal building shake. Quinn saw Karin flinch. The grenade bounced precariously on her chest. Her eyelids were pressed together, lips sucked in. Quinn realized Arbatov had blown the computers. He fought the urge to push open the door but backed away into the darkness. The thin copper wire ran to the printer knob. All Arbatov had to do was touch it.

*   *   *

In the chamber, thick smoke billowed toward the curved ceiling, tongues of flame licked the air. Circuits popped and snapped as the computers were consumed in flames. Freon, used to cool the computers, vaporized with a hiss as it came in contact with the air.

Nick and Running Fox, working together, untied the hostages, ordering them to lie flat against the concrete floor as close to the building as they could get—out of the line of any possible fire coming down from the second level. The temperature, a constant seventy degrees during normal operations, was quickly rising. The lead seals on the extinguisher system in the chamber's ceiling melted under the intense heat being generated and water began to cascade down. Nick covered Running Fox as the Ojibwa struggled with a fire hose. A hostage got to his feet and came to his aid. Soon another joined and in seconds, ice cold water from above and below, hitting white-hot metal, created balls of superheated steam. The Third level was enveloped in thick smoke and steam.

Within minutes, the CRAY Y-MP2E's were metal skeletons, some parts glowing red, crumpled under the cold water bath. The four units, worth over a hundred million dollars, looked like witch's fingers clawing at a leaden sky.

*   *   *

Feeling the change in temperature, Quinn clipped onto the open line. "GATEKEEPER to topside . . . do you read?"

"Captain KHE SAHN—here Sir!"

"BRIGHTSTAR on line!"

"ROCKAWAY on line."

"Caleb, Jocko here. Report!"

Quinn sucked in a breath of warm moist air. "Hostages should be secure but the computers are gone. I'm separated from the chamber by the depth of the service building. Arbatov blew the CRAYs as soon as the run was complete. I saw printouts but believe the tape transmit system must be on the third level. He has two hostages left and both are booby-trapped. He must feel that either he can transmit or he's going to make a run for it. What's happening up there? Has the Head Frame antenna been taken out?"

"Colonel Quinn. We're about to commence and Holmes is calling the shots."

Quinn slumped against the work desk, his eyes watching the narrow opening in the floor leading to the second level. Dull but persistent bolts of pain radiated down his arm, up into the nape of his neck.

*   *   *

Ernie Holmes had moved his observation post from the western edge of the rock pile further east. Now he had a commanding view across the pit at the Head Frame, Dry Room and Engine House. Two dark green Harriers shifted from forward to vertical flight control. They took up positions called out by the National Guard officer hidden among the pines and balsams rooted in snow-covered granite and graywacke. Riding on shimmering columns of hot exhaust gases, the two marine fighter-bombers, dropping close to the ground, sent up billowing clouds of snow and ice. Nose tilted down, one put sporadic fire into the Engine House. The second slid into position—level with the olive drab antenna located just above the assembly of the Head Frame. Major Holmes gave the order to commence firing. One short blast and the antenna turned into scrap iron.

*   *   *

Quinn unzipped his hunting parka. Moist heat outside the service building had eased its way into the confined inner-module. It was heavy with the smell of burnt plastic, paint and charred wood. The phone in his hand, a link to the surface, helped to mitigate but did not eliminate the feeling of claustrophobia.

"ROCKAWAY here. What happened? Caleb you seemed to slip away from us."

"Sorry Mel! Have you got your computer rigged for the Russian language."

"Try me," the response from Langley.

"What does the word "*pita*" mean in Russian?"

"As in *pita* bread?"

"Dunno. It was a word we heard down there. The other was coffee-just like the word except these people were drinking tea."

"The last one's easy Colonel. Coffee in Russian, transliterated comes out *K-O-F-E*. Hang on I'm zipping through the 'P' sound on the computer."

Noise from the crawl space caused him to swing around, aiming the M3 at the dark hole. Running Fox, striped face smeared, sweating, pulled himself onto the metal floor. "The computers are dead. So are four more Russians. Nick's got another Purple Heart coming and Wally worked his way back through the tunnels. He and Gramps have something in common to talk about—rock splinter wounds about the head and shoulders." Quinn saw a flicker of a smile in the Indian's eyes. He gave Jimmie a thumb's up salute as Mumford's voice filled the receiver.

"Pita. Forget bread Colonel. There's a word *P-T-I-T-A-S* in Saint Cyril's script. It means bird!"

Quinn's eyes snapped open, free hand slapping his forehead. "Bird, Mel! Coffee Bird-the satellite used to read weather over South America . . ."

Mumford cut him off. "Hang ten—there's a file coming up! Got it! Fired into orbit on July 17, 1991. Airianne launch in French Guyana— successful. Client was Callibrio Export Company, Sao Paula, Brazil. Part of a two-satellite dump. One got boosted out of orbit when a jockey jet stayed open—lost in deep space and one is operational. Weather Orbiter Number 77 in North-South orbit! It could be . . . ."

Quinn finished the sentence. " . . . a Russian Sleeper."

*   *   *

Electricity seemed to charge throughout the Situation Room as the monitored message came in over the receiver. Yoder had already recalled the global satellite map. In seconds, the estimated trajectory of Satellite WO77 was displayed. Swinging over the horizon from the northwest, clearing to the southeast, the rough-cut path was east of Tower-Soudan by a hundred miles. He picked up the phone reserved for direct contact to Cheyenne Mountain.

"I want that satellite's exact flight path and time of arrival over the northern horizon." The CIA Chief turned away from the screen to face the President. He said nothing.

Jaw muscles tight, the Chief Executive looked around the room. He could cut the "*I-told-you-so*" feeling with his gold pen. *Some grand gesture, Mr. President! Blow up a satellite to prove good faith and another one does the deed. I don't think I can hold off my people any longer.*

The President massaged his knee. For the past two hours, he'd been without pain. Now the weakened ligaments sent familiar signals. The knee was warm to his touch through the fabric. Pushing notes and pads to one side, he did not see Doctor Chen flinch visibly as his last message was buried under the papers.

Yoder put the phone down and turned to the screen. A computer-generated path appeared on the screen. Like lights on a Christmas tree firing in sequence, a corrected path was described. The trajectory was directly over Tower-Soudan.

*   *   *

"Perhaps we should think seriously about leaving the Kremlin to join the members of the presidium." Korovkin put his green flimsy on the stack of earlier Hot Line Messages. His voice was level.

Manarov held up his text; "Any submarine deemed hostile or showing hostile intent will be dealt with by whatever means available. It is a new deadline, another gauntlet thrown down!" He glanced at the Field Marshall then at the head of the Commonwealth.

The Russian leader caught the look—like a dog in the pack—sensing a battle for leadership. He adjusted his glasses. "So another satellite is to be disposed of. Is there any reason why we should be concerned about the destruction of Weather Orbiter 77?" He looked at Korovkin, then Manarov.

"Nyet," responded the KGB officer, his face impassive.

Korovkin smiled; "Nyet, Mr. President . . . destruction of still another satellite is simply a pretext—high altitude gamesmanship. What *is* of concern is having our fleets constrained, our submarines threatened. And, I must warn you, further delays will cause massive problems with China. We are beginning to see activity across the border. Should we begin war with the West, it is likely the Chinese will see this as an opportunity to begin hostilities on our Eastern borders, playing into the hands of the US by forcing us to divide our resources. Is this something you would care to explain to the presidium?"

Touche`, Field Marshall Korovkin. The President turned aside the General's question with another question; "Gentlemen, how long after the new American deadline will a major incident occur?" The Russian Leader addressed the question to both men.

Manarov responded first. "Within an hour."

"Nyet," responded Korovkin. "Within minutes!"

The President pursed his lips. He glanced at the wall clocks and back to Manarov. "Order the evacuation helicopters for those of us remaining." Sweeping his notes and papers together, including the unread message from the Dutch Ambassador asking for an urgent audience, the President took one last glance at the portraits. *The eyes,* thought the Russian Leader, *they seem focused on me.*

*   *   *

Re-entering the conference room, Eli Ben-Yaacov, handed flimsies to the Prime Minister and Rabbi. "This is like going to the executioner's scaffold on a fast escalator! Aaron cannot reach the President—not even Dr. Chen. Our unofficial go-between in Moscow, the Dutch Ambassador, informs us that it is impossible to reach anyone in Moscow of *any* rank, much less the President. Our conduits to White House and Kremlin are shut down."

The Prime Minister, hand trembling slightly, put his copy of the U.S. transmission down, eyes focusing on the Rabbi. When he spoke, his voice was almost as soft as the wisp of the ceiling fan. "Rachman's Dove is our last . . . ."

General Yaacov, anticipating long hours strapped in his *Kfir,* remained standing and nodded in agreement. "It has come to that—we

250

will execute the Rabbi's plan . . . may it work . . . *Eem Yeertzeh Ha-Shem* . . . ."

"God willing," repeated the Prime Minister.

*   *   *

"Colonel Kiger!" The American sergeant called for the watch officer seconds after the message had begun. Standing behind him, the colonel watched the computer printout roll upwards. He glanced at the clocks and shook his head. "It's not possible sergeant, just not possible. Nobody enters this system and just starts talking to the President of the United States and Russia!"

Acknowledged by his peers and commanders as a "*techie*" of the first order, the sergeant nodded. "Somebody *has* and *is* Colonel! If I had to stick a pin in the donkey's behind, I'd take a peek at the land line in Tangier."

Colonel Kiger, half hearing the sergeant, his mind rocking at the implications of the message reached over, tearing off the sheet. In seconds he was on the phone.

*   *   *

Illya Lukavonovich, former member in good standing of the Communist Party and a Major in the KGB, watched the unannounced message appear in clear Cyrillic letters. Lukavonovich realized, a third party had tapped into the sacrosanct communications system. Hurriedly tearing off the completed transmission, she reached for the foam-green phone to the right of the printer.

*   *   *

"The fire's out, Colonel. LaMont's people are quieting down. Some are in bad shape emotionally but otherwise ok. Everybody wants out of the mine. The second deck is empty. Whatever came out of the printers is gone. No Arbatov and no prisoners—he must be holed-up on the third deck getting ready to transmit on the t ack box."

Quinn gave a short, sardonic laugh then reached out to Nick, his hand gently lifting the old Finn's stubbled, blood stained chin. "You're gonna need some medical help here, Nick. Rocks, lead, bat shit."

He picked up the phone. "GATEKEEPER . . . who's with me . . . what's happening topside?"

The litany was repeated.

"Caleb—Henderson here. You're last bit of information has the propane burners firing again—the balloon is straining to go!"

"What's the program with the coffee-bird satellite?" asked Quinn rubbing his face. His mouth tasted of burnt cotton.

"The only package available is the matched pair of ASAT's on the deck in Texas. They're scrambling again to catch it on the next pass—coming in twenty minutes."

"Too far south General! If that black box is rigged for encrypted bursts and they miss it it's over at this end. He could encapsulate the data and give Moscow the key elements in a single microburst!"

"Understand that, Caleb. But remember—Arbatov is down to one antenna. We nail it and it's over for him!"

"Quinn! Colonel Quinn, Marine!" The sound of the flat voice echoed throughout the shotcrete lined cavern. It had the metallic ring of being amplified. Nick pointed upwards with the barrel of his rifle. "Jahah! That sonofabitch'n Russian's got your number!"

"Jocko, Arbatov wants to talk! I'll try again down here!" Handing the phone to Nick, he spoke rapidly; "I'm taking Jimmie with me to the front of the building. Keep the line open!" Snapping a full clip into his M3, he moved through the cafeteria past the corpses of two Russians. Running Fox followed. At the doors, he stopped, pushing them open but remaining within the building.

Loose wires still sparked ahead. Tendrils of smoke rose in the dank air. A layer of haze, the color of phlegm hung just above the charred remains of the computers.

"*Dante's Inferno*," said Running Fox quietly, "*last level down*."

Rubbing his eyes, Quinn looked at the Qjibwa for a fleeting moment and back to the charred ruins.

"Quinn" the voice boomed again.

Edging closer, he looked up. There was no way the Russian could fire down. He scanned the chamber. Hostages crouched under the building itself—momentarily out of harm's way.

Cupping his mouth with his right hand, he answered back: "You've got Quinn. Talk!"

The amplified voice filled the chamber; "You're to be commended Colonel Quinn. With nothing but a ragtag handful of men, you've done

252

well. I, however, have professional soldiers—and your friends as hostages. Interfere and they die!"

Quinn looked at Running Fox, then upwards. "Chances for you and your remaining men getting out alive are non-existent, Colonel Leonid Arbatov. Give up. End this bloodletting!" Eyes burning from the smoky residue, he wondered if the upper floor was sealed. *Arbatov and the women could be in worse shape up there than we are down here.*

Arbatov's voice boomed again: "Concern yourself with your own life, Colonel! There is enough plastique left to bring this entire complex down!"

Quinn heard an audible click. He turned to Running Fox. "Our Russian chief owns the moment. Let's get back to the phone!"

They worked their way quickly back to the communications link.

"You've got'em Colonel. KHE SAHN on line. Hell of a show going on up here! "The command track's still a big zero but it's blue skies. Two degrees above zero and minus forty-two wind-chill! The tanks look like ice palaces. Some of the Russians tried to hold off the bombers but the Harriers laid down suppression fire and they dropped back into the hatches. Ice formed in seconds from the big drop. The helicopter is relaying water over the hill from Stuntz Bay for the finishing touches. The tanks are dead! If they fire, they're likely to rupture the cannon.

"Hey! The command track just popped up—broke out of a small outbuilding tucked next to the Super's house. It's coming toward the Head Frame antenna scanning! I can take it out with a Harrier! Orders Colonel!"

The raspy voice of Henderson filled the set.

"Take the Command Track out Major Holmes! Now!"

Quinn screamed back. "BELAY THAT ORDER! THIS IS COLONEL QUINN ERNIE! REPEAT! BELAY THAT LAST ORDER!"

He could see his former Recon man huddled in the rocks above the battlefield, torn between his senior officer in the field and the Commandant of the Marine Corps in the same room with the Commander-in-Chief of all U.S. Forces.

"Ernie, if you order Harriers to ace that APC, Arbatov is likely to lash out at the rest of us down here. He's got plastique that can turn his place into a tomb! Let the sonofabitch go!"

"Caleb, this is Henderson. If we take out the command track this thing is over! It's a calculated risk. If Arbatov can't transmit, he'll probably give up!"

"Joc—sorry—General Henderson!" I can't buy *probably*. This man's a professional killer! I've seen him operate. This is a suicide mission. If he can't succeed he doesn't give a damn how many people die. If you order that APC taken out you kill me and thirty others. I can do something to keep these people alive down here! LET THE RUSSIANS HAVE THE APC!"

"Colonel Quinn, this is the President."

Eyes closing, he braced himself physically and mentally. *Your country needs you. Your country is depending upon you. Everyone must make a sacrifice. BULLSHIT!*

"Colonel Quinn, you've done one hell of a job! We realize you and your people aren't out of it yet. I've just informed your boss that the APC will *not* be destroyed. It's my understanding that you have a clear picture of what DEEP ORCHARD is all about. Is that correct?"

"Yes Sir!" Quinn wondered what the President was leading up to.

"There are developments I can't discuss over this circuit but you need to know—it's quite possible the men you've been up against are, in fact, SPETSNAZ but are not doing this for Mother Russia. This won't change the complexion of things down in the mine but it's having a significant impact on the surface—the entire surface of the world. Confusion is understandable. So do what you can to stop the transmission at your end. Protect your people. We'll do the best we can."

A sharp detonation rocked the small room. The metal walls surrounding the three men vibrated. The phone died in his hand.

Jimmie's carbine snapped into the air. "Grenades. Fired on the third level!"

* * *

The President and his immediate staff watched the activity from the far end of the room. Each military Chief of Staff was issuing identical orders to their respective commands—freezing all operations dead in their tracks. No going forward—but no going back.

The President turned in his chair to face Dr. Chen: "What do you make of the Israeli move?"

"Ah! A stunning display of diplomatic chutzpah! I know the Prime Minister. He must have bitten the bullet before agreeing to expose his penetration of the Hot Line."

"Your thoughts Tom?"

A thin smile creased the NSA's face. "The Israelis didn't have to explain how they penetrated the system—the very act was demonstration enough. They gave up plenty by exposing their capability. It's plain to see, however, that they had everything to lose by holding back."

"And the Penitents as described by this Rabbi Rachman – Doctor Chen, what are your thoughts?" The President looked at the small man with the big brief case still at his side.

"Ah! An embarrassment to the Russians who pride themselves on internal control! To have penetrated the highest ranks of the Soviet Military as far back as the 1940's—it is long time for a mole to burrow. Heads will roll and the upheaval might serve to insure world peace longer than rocket size or numbers. As to the Penitents themselves, they've survived for thousands of years. This will probably be only a temporary setback in their scheme of things." The little Professor snorted. "I think they will find some other way to manifest their penitence in the future, however, thanks to our Colonel Quinn."

The President nodded in agreement. "But if this Colonel Arbatov succeeds in transmitting the product of DANTE'S ORCHARD, their people—somewhere will still own an asset beyond monetary value." They'll review the data, find a spot relatively unaffected by the Polar Shift and relocate there, bide their time, propagate the faith. They'll beget, as the Bible says, and be heirs to the post Polar Shift world." He paused and shook his head. "You've got to give them the piper's due. The *Penitents* could be the one and only religion."

Matteo Chen laughed softly, white eye brows twitching, "It is ironic . . . a religious sect borne of Judaic roots and the Diaspora, forged on the anvil of enforced conversion, then melded to Catholic and Islamic tenets now comes to the beginning of the third millennium

255

and perhaps a new world. No stone slab, no gold plates or Torah for the *Penitents*, their survival in the valley of tears made possible by a— computer printout . . . ."

*   *   *

"Welcome to the Green Room, Colonel General Bogodyash, and Mr. Dunayev." Yuri Trishenko motioned to the two men, to take the chairs recently occupied by Korovkin and Manarov. The baize tabletop was cleared of papers.

"General Bogodyash, I passed over twenty men senior in rank to make you a Field Marshall of the Russian Army." He turned to the man called Dunayev. "In your case, Stanislav, as an outsider and an academic, you will bring a fresh approach to the business of intelligence gathering in these changing times. I expect great things from both of you. But now we still have a major problem." He looked directly at the young officer who'd been Russia's third man in space and, until his selection by Trishenko, headed Russia's air command. Using the rosewood pen for emphasis, the President tapped the tabletop. "Can we destroy Satellite 77?   Can it be done before it reaches this *'window of reception'* the Americans talk about?"

The young officer nodded. "With more time, we could explore a half-dozen options. As it is, we have only two. I had my people at *Kalingrad* and *Tyuratam* work the problem. From *Tyuratam*, within the next twenty minutes, we can use laser to render the unit inoperable. The risk is that there is no assurance of success until the unit is asked to receive signals. Over that we have no control. Here is a back-up option Mr. President; *Kalingrad* reports that *Mir* will be in a bisecting orbit in one hundred and ten minutes—directly over the United States. The Americans seem confident enough of taking the unit out somewhere over the southern part of their country, but that may be too late. If we make the decision now we can kill the bird twice. The laser at *Tyuratam* and the bird itself over the U.S. This should be a *'good faith'* display on our part   . . . there is a price to pay, however . . . ."

Looking over the top of his rimmed glasses, the President waited for clarification.

"The Americans do not know we have such powerful Excimer lasers aboard *Mir*. We use them only between the spacecraft and test targets in deep space. The moment we trigger the beam over their

256

country, they will know we can destroy virtually anything in space. It is, Mr. President, a secret of great importance to the defense of Russia— our only *ace-in-the-hole*—to use an American term, against their very potent Star War Defense system."

The President had the evidence that confirmed his suspicions about Korovkin and the military leadership. *They too, had something up their sleeve.. . and now—by God . . . they will be shirtless! . . . .*

The round-faced Dunayev laughed, raising his hand. The President and the new Field Marshall looked at the smiling Professor.

"Two countries have already exposed closely held secrets—the United States, that it has this installation buried underground in a remote province called Minnesota, and Israel's disclosure they have been listening to our supposedly '*secure*' communications for some time. We have here an opportunity for reciprocity in a matching manner. I say we press the button on board *Mir*—help the Americans—and perhaps gain a bargaining chip when we meet with the bankers of the world—as surely we must . . . ."

Twisting the rosewood pen slowly in his hands, the President smiled ruefully at the astute reasoning of his newly chosen KGB chief. He looked at the pen a moment, then up to the clocks. The smile began to spread over his face, his head bobbing. He pointed his pen at the newly boarded Field Marshall. "Order it done! *Tyuratam* fires its laser and *Mir* provides the *coup de grace*! Inform the Americans immediately!"

*　　　*　　　*

"Nick, we've got to get the people out of here. There's no telling what Arbatov will pull next." The sound of heavy movement filtered down through the service module from above. Running Fox, in response to the explosion, had been ordered to the second level. Communications cut off, Quinn put the dead handphone down.

Nick snapped his fingers. "Jahah! I damned near forgot! Down the tunnel about fifty yards there's a place that holds up to seventy-five people in case something happens in the mine. Gots auxiliary power, food and heaters. First aid stuff, too!"

Gripping Nick's shoulder, Quinn said to the old Finn, "You and Wally are in charge of the hostages. Get them moving outta'the chamber and into the safety area. Keep away from the front of the

257

building and don't give our Russians a clean shot. If I read the clock right, he must have his hands full getting ready to transmit."

Nick nodded. He gave Quinn a thumbs-up sign and disappeared down into the crawl space. Looking up at the open hatchway separating the lower level from the computer room, Quinn saw a bulky shape edging downward. Weapon pointing upward, he pulled his trigger finger back.

Running Fox, breathing heavily, was carrying his sister. Reaching up, Quinn took the girl's lifeless form and gently laid her on the floor. Warm, sticky fluid oozed onto his hands. Playing his light over the girl's dark features, he saw slight movement of the nostrils.

"Shrapnel in the back Colonel. She and Karin decided one of them should make a move. Val said she got her hands free and the grenade off her neck then threw it over the Russians shoulder toward the corner of the room. He ducked, rolled away and she fell near to floor, her back to the blast. Karin was protected by the mattresses."

Quinn looked at Running Fox. "Take her down into the crawl space. Nick and Wally are going to move the people out of the chamber and into a safety room. There should be some medical supplies. Stay with her, Jimmie."

"She comes from tough stock, Colonel. You help me get into the crawl space. Nick and Wally can take her from there. You're gonna need help!"

Quinn nodded. Working together, they lowered the young woman into the narrow crawl space. Saatela was there on his knees to receive his sister-in-law's bloody form.

Extending his arm into the hole, Quinn shook hands with the deputy. "Nice going Wally!" Bloody, his eyes were glazed but smiling.

"I got one, but there's probably one left at the bottom of the hoist waiting for his boss. Gramps is leading the people to the safety room."

"Good, now help Jimmie get his sister into the Safety Bay and head back here."

Alone, Quinn looked up at the open hatch above him. His back against the wall, he slowly slid to a sitting position, eyes closing. *Karin's still a hostage and Arbatov has plastique—lots of it!* Head tilted back, he forced himself to think about the enemy above.

"Damn!" Uttering the word, his eyes snapped open. "Booby traps. That sonofabitch is going to use the plastique, one way or another!" The air around him was warm, still smelling of things burnt. The metal floor under his buttocks was cool but not yet cold. He clenched his trigger hand. The fingers moved slowly but still moved. He knew he should get up, stay active but the dark room and warm air had a near narcotic effect. Layers of dried sweat crusted between skin and thermal underwear. Scratching the black stubble on his chin, thoughts of steaming hot coffee, eggs over easy and crisp hashbrowns flashed into his mind. He could smell the dark coffee, almost savor its taste. One fantasy bred another—the tantalizing vision of the meaty Finnish stew, Karin's laughing blue eyes looking at him over the rising fragrance, handing him a chunk of warm, moist bread, telling him it was okay to dip. Why would anyone want to leave her?

They finished the meal, enjoying each other's company. Karin walked down the hill with him to the shore of Beatrice Lake. Taking a leisurely sauna, the towel dropped slowly from her small but well-formed breasts, coming to rest in the "V" of her slim legs. She cradled his head and drew it down. Lips barely touching, she gently pushed his head down further. His lips touched the olive drab grenade nestled between her breasts. Eyes snapping open, he pulled back in revulsion.

Rivulets of sweat ran down his back. Staring into the dark, wondering how long he'd been sitting, the need to relieve him self was urgent. Standing up, fumbling with his fly, he tried to erase the image, then let his bladder go, the stream running down the wall, odor rising upwards and masking his memories. Zipping his canvas hunting pants, he made ready to turn but the room exploded from above in a burst of gunfire. Metal flew everywhere and Quinn instinctively dropped into the corner, hands over his head. The noise was deafening and shards of hot metal peppered his exposed arm, side and thigh. The smell of his own urine, overpowering.

Metal seemed to move everywhere. Then there was a brief pause before another prolonged burst. *He's using full clips to clear the area. He's either made the transmission or found out he can't.*

*   *   *

Anna Gritsenko, Computer Specialist, dressed in a dark green jump suit, glanced at the large tracking screen alive with Russian and American targets. Checking the position of Kosmos 1124, a test bed satellite that looked like a giant Christmas tree ornament, she made some last minute adjustments on her computer. Outside of the blast-hardened facility, in clear, cold air, a large, three-story gimbaled gray metal device responded smoothly to her commands. The characters on her screen confirmed that the Laser Gun had achieved Target Lock. Pushing back from the console, Gritsenko flipped the firing toggle. Ten seconds later, light amplified beams pulsed skyward making direct contact with a mirrored plate on 1124 which in turn was deflected like a billiard shot toward the 270 pound commercial satellite known as Weather Orbiter 77. The high intensity beam made a pinhead size hole in the titanium outer shield. It was sufficient. Reaching for her phone, Anna announced quietly, *"The target has been painted."*

*　　　*　　　*

Wearing a maroon sweatshirt from the Massachusetts Institute of Technology, souvenir of the Appolo-Solyut space mission, Colonel Boris Tagashov, seated in the command chair of the *Mir* space vehicle, pulled at the rubber exercise handle. He'd already been seven months in space, and found himself thinking more and more frequently about a shopping trip to *Valuitnyie Magaziny,* the store off Gorky Street still reserved for Heroes of the State like himself and other top ranking members of the Government. Thoughts of caviar, choice beef and American bourbon filtered through his mind.

Attitude thrust jets had been employed to bring *Mir's* viewing ports in line with the trajectory of WO77 as it crossed the space lab's bow on a flight path thirty-seven miles distant from the orbiting laboratory. Tagashov put the rubber device under Velcro control at the right of his chair and began the laser firing sequence. He looked out the viewing port. Great Bear Lake in Canada's Northwest Territories was in his field of view. There was a black object against a white background on the target screen. He hit the last switch. The laser cannon fired, its jade green eye blinking twenty-two hundred times in the space of one second. WO77, already electronically dead, blew apart under the sustained bombardment. Panels separated and the main frame collapsed into the camera and radio chamber over Prince of Wales Island. Seconds later, Belcher Island residents in Hudson Bay saw their second fireworks display within eight hours. Five minutes later, CELESTIAL

260

EAGLES III and IV, circling in blue Texas skies, received their stand-down orders—ASAT's still hung on hard points.

* * *

Yuri Trishenko pressed a small button on his walnut desk. The door opened and Dunayev and Bogodyash turned as he directed their attention to the small, dark-haired man with a deeply lined face who was escorted into the room by a young army officer. Rising from his chair, the Russian Leader came from behind his desk and approached the elderly, black-suited man whose piercing eyes where a match for those of the President. Extending his hand, he grasped the man's hand and forearm, nodded with a smile and then turned to his handpicked replacements. "Gentlemen, I would like you to meet a fellow Ukrainian, survivor of the holocaust, our own Gulag and a priest of a sect we need to be more knowledgeable about. *Alexsandr Binyaminov*, please, be seated. There is much to discuss."

* * *

Minutes after the shooting had stopped, Quinn's volunteers surfaced from the crawl space. Running Fox first, Wally next. The grandson and Quinn helped the old Finn out of the hole.

Quinn pointed to the ceiling and the numerous bullet holes. "Our Russian friend wanted to keep us honest. He must be planning to get out with Karin as hostage and whatever came out of the computers. We've got to get back to the main shaft!"

Nick started for the metal stairs but Quinn intervened. "If we go in his path we'll get blown to kingdom come! He must have taken the route out we took in. It won't take much to string trip wire grenades behind him. We'll have to get to the cage on this level!"

* * *

Ernie Holmes, from his icy command post in the rocks above the Engine House, surveyed the panoramic early afternoon scene. In every direction, white clad troops on skis closed in on the mine site. A Chinook, in desert colors, landed in a snow-filled gravel pit and men were already moving to take up positions on his flanks. A marine lance corporal sighted in on the Engine House with a shoulder-fired rocket. To the south, the last of the Minnesota Air Guard C130's unloaded troops on the highway and were airborne in azure skies. Harrier jets, four of them on the ground beyond the frozen tanks, waited for the call

261

for air-to-ground firepower. Twisting his binoculars to the maximum, he could make out the bulky shape of BRIGHTSTAR, the plane linking him to the White House, on station five miles above. Every few minutes, a Strike Eagle would make a screaming, low-level pass over the Engine House to remind the Russians still alive that the US military had arrived in full force. The wind, now blowing from the North, picked up an occasional puff of white snow and dusted it over the landscape. Vapor from hundreds of soldiers laboring through waist-high snow was whisked away in the icy cold. The Forward Air Controller had nothing to report except the wind-chill factor, which was 40 degrees below zero and falling.

*   *   *

Quinn stopped twenty yards from the base of Shaft Eight. He turned off his lamp and looked into the dark cage. A voice, speaking Russian echoed near the top of the unit. Jimmie crouched on the tunnel floor. Quinn motioned Nick to move up, whispering to the old Finn; "They came down in the top cage. Will they go up that way?"

"Jahah, the phone's in the top unit. He calls the Engine Room from there."

Reaching for the small transceiver still hooked to his inner jacket Quinn flipped it on. A low level of static gurgled in the set. He flipped it off.

"Nick, we came down on top. we can't make it into the lower unit without his knowing we're there. He's likely to sabotage the lift mechanism as soon as he gets to the surface. Have we got any options?" Turning his flashlight back on, the orange-black iron seemed to absorb light. Nick shook his head. "It's a damned long climb up. We hooks a safety cable under our arms and starts up. There be a rest platform every fifty feet. You younger troops could make it but this old soldier couldn't."

Quinn let the beam of light fall between the narrow space directly beneath the skip. He saw more rusty iron. "What's that under the cage?" he asked in a low voice.

Nick stared in the dark, his eyes brightened. "Jahah! We have to give them eggheads their due! It's a skip they left hooked so's they could haul more gear up and down!"

262

In the lead, Quinn worked his way closer to the shaft. Voices above grew louder. A dim light cast shadows around the iron cage. Quinn recognized Arbatov's monotone. Already in the upper unit, he guessed the white-haired colonel was ordering the cage to be lifted.

Unable to communicate with Holmes—to stop the lift mechanism—he racked his brain for a solution to their problem.

Stepping forward, he pointed the flashlight into the dark space beneath the bottom cage. The rust-covered skip was on track, hooked to the lower unit. It was big enough to hold all of them. He moved forward, turning slightly to motion Nick and Wally up when he felt something brush his shoulder. It was a peach-sized grenade wedged at eye level. A thin wire, circular arming pin clung to his elbow.

He batted the armed explosive on the rock floor between himself and shaft. Spinning, dropping and forcing Nick and Wally to the tunnel floor in one motion, he felt Running Fox surge past, black snowmobile suit a blur in the dark tunnel.

Instinctively, Quinn opened his mouth. The explosion hammered his eardrums as small shards of hot metal ricocheted off the tunnel wall, peppering his back. Cordite burned in his nostrils and eyes and a ringing sensation filled his ears. The voices above were louder, animated. He pushed up against the rock wall, Nick was unconscious, Wally, eyes blank, was bleeding from his nose, mouth and ears. He turned toward the base of Shaft 8 and saw the still form of Jimmie Running Fox. Eyes swelling with tears, he knelt beside the Ojibway and carefully rolled him over. His chest cavity was crushed, rock floor a pool of blood. The young Indian's face was serene. *"Libero nos Domine,"* he murmured as bright red blood bubbled from the corner of his mouth. The eyes closed.

Quinn held his grease gun in the crook of his arm and fired an entire clip at the base of the cage. Bullets splattered off the walls and ricocheted in the lifting chamber. Sound of the useless fusillade echoed throughout the drift. Quinn closed his eyes for several moments in silent prayer. When he opened them he saw Wally staring at his *niita's* corpse. "Take care of your Grandfather," he said softly. "Jimmie's dead. I'll get the cage back down as quickly as I can for you and the hostages—somehow—I promise."

Wally nodded in dumb silence.

263

Quinn heard the three-burst ring signaling lift was imminent and saw the familiar snap of the cable. He stood at the edge of the shaft as the cage jerked once, then began to rise. Remembering how quickly it had accelerated downward, he judged the distance, put his hands behind him, crouched and pushed off, grabbing at the rim of the skip.

The M3 banged against the rusting iron side but the sound was lost in the whir of cable running through the thimble. The field radio hung loosely from the other shoulder.

Hot pain shot through his body as torn muscles were stressed again. The cage hurtled skyward. Quinn dug into the scaly iron, feet searching for a ribbed part of the ore cart. Iron wheels on iron track clattered next to his head. Closing his eyes, he hung on. Groping with his left foot, he found another iron rib. Pushing up relieved the pressure on his upper body. He was able to put his good arm over the top of the skip.

His right foot found another narrow indentation. Quinn willed himself upward. He let the M3 slide off his arm, catching it by the sling and swung it into the mouth of the skip with a clatter. Another burst of pain shot down his spinal cord. His head felt like it was being torn from his body. Taking a deep breath, he pushed again. Fighting gravity with ebbing strength he pulled his upper torso onto the lip. Balancing precariously, he finally plunged headfirst into the ore transport, the radio slipping off his other shoulder but staying inside the skip.

*   *   *

Holmes saw the cable tighten before he heard the sound coming out of the windowless Engine House. His eyes followed the twisted cable as it came off eight-foot drums inside the Engine House. Then he heard a scraping sound slowly building in intensity. His gaze swinging from Engine House to Head Frame he saw the source. Olive green metal, twisted remnants of the transmitting antenna blown away by the Harrier's machine guns were caught in the main cable assembly and were being drawn into the sheave itself. The lone Russian in the Engine Room, unable to see the problem and apparently unfamiliar with gauges indicating unusual stress build-up left the lift levers set at maximum power. Holmes watched the cable slide through the jumbled mass of iron and heard the grinding sound turn into a high shriek. Smoke appeared as metal abraded metal. Individual strands of the

264

three-inch thick cable began to melt under the intense pressure. Now at a screaming pitch, the cable began to disintegrate rapidly. Hundreds of white-clad marines, Norwegian counterparts and repatriated Guardsman, ten yards apart, advancing toward the buildings, stopped in their tracks. They also heard the high pitched whine. The final parting of the massive wire sounded like the crack of a gigantic bullwhip. It echoed between buildings and over the rock formations. Holmes ducked instinctively as the huge cable whipped back over the Engine House. The giant cat O'nine tails sliced the tops off ancient pines over his head. A shower of pine needles fell, carpeting the snow-covered ground.

*   *   *

Quinn heard the unearthly sound from the bottom of the skip. For several seconds the cage was motionless. Then he heard a violent snap. The cage jolted downward a few inches. After a second snap the cage and skip dropped like a rock. Karin's high-pitched scream was quickly drowned out by a sound like a giant buzz saw cutting through steel rails. Quinn covered his ears, gorge rising in his throat. If the cable had parted, he was seconds from being crushed to death. Unable to think or to react, he bounced violently about in the ore skip. Then he became aware the cage was slowing, then stopped. The smell of burning wood filled the air. He heard Arbatov's voice, shrill, demanding.

Moving to his feet, Quinn found his flashlight and looked above his head. The cage was wedged between the large timbers by a series of wooden pegs that spread apart under the pressure of springs. He shook his head weakly. *Nick was right. The one hundred year-old safety system worked as advertised . . . .*

*   *   *

Spinning in its track, the whitewashed APC started to move up the incline toward the Head Frame, dish antenna rotating from side to side in a short arc.

"KHE SAHN reporting in. APC appears to be moving to support whatever troops are left holding the Head Frame. Cables have parted. We don't know what happen to the cages!"

"Henderson to Major Holmes. Order the infantry to hold fire. We want prisoners and information. Not dead Russians!"

265

Ernie Holmes relayed the order from the Commandant to the ground forces closing the ring. In the distance, white clad troops were crawling up on the immobilized battle tanks. The troopers worked their way to the base of the Engine Room. A single shot rang out from inside.

* * *

The cage was wedged to a stop by the safety system fifty yards from the top of the shaft. Quinn moved up the ladder half the distance and came level with the base of the top cage, M3 slung over one shoulder, radio the other, he eased toward the open door. As his head came up over the floor of the cage he saw Karin—on her knees.

"Don't come any closer, Caleb." Her voice was a hoarse whisper. Hands still tied behind her back, the olive drab grenade stood out against the white, now begrimed Riverside sweatshirt. A six-pointed gold device dangled under the grenade. It was attached to the arming pin by a bent paper clip.

Laying weapon and radio gently on the floor, Quinn removed the flashlight. Karin's blue eyes were wide, face stained, with blonde hair hanging in damp, twisted strings. "The pin Caleb, it's barely in." Her voice was low as if speaking louder would give added weight to the religious device now a macabre trip pin.

Something bounced on top of cage. Seconds later an explosion rocked it. Landing on top of the lift, they were shielded from the white-hot shards of Arbatov's grenades. Looking up, he saw the dark form twist, a gun barrel swing out and down.

Moving quickly, he pushed into the lift. It moved slightly. Karin moaned. An instant later, the top of the cage was struck by the sustained burst of a full clip fired from above. On his knees like Karin, facing her, they remained frozen as .45 caliber rounds ricocheted off the top of the cage, chipping granite and splintering wood beams. The reberverating cacophony was like nothing Quinn had ever heard in combat. When it ceased, he looked up and saw Karin, eyes shut, lips pressed together, her body shaking.

He had to end her prolonged terror.

"Karin, spread your knees apart. Slowly, just an inch or two." he ordered quietly, "I'm gonna'put the flashlight between them—facing up—I need both hands free!"

266

Silent, Karin obeyed and a shaft of light illuminated her pale skin. For one brief moment, Quinn stared transfixed, he had seen this painting before . . . Madonna in The Tomb . . . he shook the image away. He could see the upper lip tremble and two tears appear in the corners of her tightly closed eyes. Edging closer, he stripped off his thermal glove and blew on the fingers of his left hand for warmth. "Hold your breath and don't move."

He saw her pull her shoulders together, her neck shrinking into the white folds of her sweatshirt in an attempt to remain immobile. He had already reached a conclusion on the grenades disposal—if he could keep the arming bar in place. He reached for the device. It was a cross contained within a six-tipped star. A crescent moon rested on the tips of the cross bar where a man's hands would be nailed. Cold to the touch, he gripped the base, lifting it slowly, relieving pressure on the paperclip arming ring. Sucking in his breath, he quickly clamped his fist around the grenade and arming bar as the pin fell into the folds of Karin's sweatshirt. In one smooth motion, Quinn dropped the grenade into the shaft knowing it would drop less than fifty yards before exploding. He gripped Karin's shoulders pulling her into a protective hug. The falling projectile exploded in the shaft sending a pressure blast up and down the rock tube. The cage shook but held firm to the ancient wooden pegs holding it in place.

"Okay, Karin, you're safe."

Her blue eyes opened. She stared at Quinn as if seeing him for the first time, her gaze dropped to her breast. She no longer felt the weight of the grenade.

"I'm sorry, Caleb, I couldn't help it but I wet my pants!" Laughing Quinn lifted her to her feet, tears of joy and deliverance streaming down her face.

Pulling the *puka* from it's hiding place, he cut the bonds. "You've had one hellava'hunting trip Lady *Lotta* . . . ."

She grasped Quinn in her arms, holding him close.

Then she looked up. "Valerie is she all right?"

"Valerie Running Fox is alive—her brother's dead. He fell on a hand grenade to save Nick, Wally and me from buying the farm . . . ."

Karin shook her head in disbelief.

267

Quinn looked skyward. "This mine cage isn't going anywhere. The cables must have broken or got shot to hell. Once the shooting stops, tanks can hook up and we can get everybody out on one trip. But there are still Russians on the loose."

Karin looked at Quinn, her arms still around his chest. "And he's got every thing the computer had to produce. He couldn't send anything from the tapes on the device he had on the third floor. The printout's packed in a duffel bag. He going to run the tape from that Armored Personnel Carrier—and he's got one bag full of explosives."

Quinn gripped Karin by the shoulders. "Arbatov told you this?"

Karin nodded.

"Did he tell you what the information was all about?"

She shook her head from side-to-side.

"What did he tell you?" Quinn looked into Karin's eyes.

"That the information was worth the sacrifice. He didn't expect that anyone like you would show up and give him the battle you did. He said several times that you must be a dedicated, professional soldier."

Quinn raised his eyes skyward again and then back to Karin.

"He's at the top of the shaft. And from what you've said, he must believe he can still communicate via satellite."

Karin shook her head. "He said it didn't matter, Caleb. He said he would get the word out."

Quinn's mind raced. He turned toward the iron ladder then stopped in his tracks. "Did he say anything about a time frame?"

Karin stared at Quinn, a quizzical look in her eyes. "Time frame? What do you mean?"

"Just that. Computers down there crunched out billions of calculations just to arrive at a time—minutes—hours—days when a certain event *would* or *could* take place. Did he say anything about that?"

Karin shook her head. She looked down as if she'd failed.

Quinn gently lifted her chin with his fingers, realizing he was riding roughshod over a woman who had been seconds away from a violent death. "What *did* he say?" His voice was gentler.

"When the computers stopped running, he looked at the last sheets and spoke in Russian. He saw me looking at him, smiled and repeated it again." She looked down and speaking slowly, sounding out the words. Quinn listened, repeated the words and committed them to memory.

*   *   *

The Secretary of State and National Security Adviser drafted the text, which the President would deliver at the re-scheduled 6:00 PM news conference. Nothing would be said of the Hot Line, near-nuclear alert or the Israeli contribution. Cooperation between the United States and Russia in thwarting a small group of *"terrorist extremists"* would be stressed. The word SPETSNAZ was deleted from the first draft along with any references indicating religious affiliation or belief. At four-twenty the President returned from his private quarters refreshed. He told his wife that a major, his wife and two young sons would be added to the White House Staff Christmas party list. When the draft was ready to be sent to Moscow for review, he spontaneously extended an invitation to the Russian President to visit Camp David when the lindens would be in bloom.

*   *   *

Yuri Trishenko offered his two new appointees a choice of *starka*, choice vodka distilled from Crimean apple and pears to which a touch of Cognac and port had been added or *Russkaya*—crystal clear vodka filtered through charcoal and quartz. Unsure of Green Room protocol, both chose *Starka*. The President poured for all of them.

"*Dhastrovya*!" In succession they toasted each other and the three oil-and-canvas men on the wall. The second toast was to the general amnesty agreed upon for all *Pokutnyky*. There would be no Gulag or other form of recrimination. It was then that Yuri Trishenko disclosed his intention of inviting the American President to visit Russia when the lindens in Moscow would be in full bloom.

*   *   *

Not sure where Arbatov was, Quinn stopped climbing at the Second Level, taking refuge in the first drift down. Karin huddled behind him.

Protected from gunfire, he stuck the antenna into the Main Shaft. "KHE SAHN?"

"This is one very cold KHE SAHN!"

"Ernie, We've got a jammed cage two levels down from the surface. Get tanks fueled and engineers brought up. Re-string the cable and get them lashed to the tanks. Have someone get on the phone to the Safety Room and have the people move onto the bottom cage, both levels. As quickly as possible, I want them out, dead, wounded and ambulatory!"

"I understand the orders. We can proceed with tank readiness but the Head Frame area is still under Arbatov's control."

Quinn thought a moment then replied, "Get the tanks ready to go! Have them move down the rail bed under the Crusher Building heading east then switch back and come down the main road toward the Head Frame. They'll be out from under his guns and ready to go when this is over. It's Arbatov's turn on the anvil now and I mean to work him over. If I read his mind correctly, he wants to get to high ground—probably your position or close to it. Cut him some slack and let him get up on that outcropping with a clear shot to the Northwest."

"Understand, cutting slack for the APC—will fade left—let him have the high ground!"

Quinn paused. "Jocko? Are you there?"

"You've got me, Caleb. It's good to hear your voice. What's the status of the computer information?"

Looking up at the opening, he could see clear blue skies beginning to turn purple. The sight of daylight and the end of his time in DANTE'S ORCHARD filled him with a sense of relief.

"Caleb! Can you read?" Henderson's gruff voice shook him out of his reverie. "I'm here Jocko! Our Russian Commander has got the tapes and LMRP in his possession—not mention beau coup plastique! Can he transmit to the Coffee bird?"

"Not possible, Caleb! It was hosed by the Russians less than an hour ago."

Quinn felt the pressure mounting to capture Arbatov alive and recover the data but also realized that if he told Jocko what he thought was on Arbatov's mind—another conflict would erupt. The order to take out the APC would follow—his countermanding order to hold fire would be overridden. Any chance to learn what came out of the computer would go up in gunfire. "Hang on Jocko." He let the field phone hang by his side, rubbing his brow.

Karin got up from the floor of the tunnel. "Are you all right, Caleb? You look sick or worried—or both!"

270

"Right on both counts!" He managed a weak grin and put an arm around her, his prickly beard against her soft skin.

It doesn't matter if the Coffee Bird is dead. This guy has a sophisticated radio left. What else could he transmit to? . . . .

He mentally pictured the surface as Nick had sketched it. Then he remembered the radio towers marked by the Old Finn—and described again by Holmes.

Looking into Karin's eyes he pulled her even tighter as Arbatov's final gambit became clear. Releasing her from his grip, he tilted the antenna back into the Main Shaft. "Sorry, Jocko I had to do some thinking."

"Understand Caleb but while you're thinking keep in mind we've got to get the printout and tapes. With the computer system destroyed and one-of-a-kind software with it, we need the information badly."

"Colonel! KHE SAHN here! The command track is moving away from the Head Frame toward the ridge line!"

Quinn smiled. *Snow Top's heading for the high ground. We can get out. Our troops take over the Head Frame and Engine House.* "Let him get to where he wants to go, Ernie, but call in the fire bombers and chopper again. Hose' him down good! As soon as he clears this area, get those tanks into position to lift the cage!"

Holmes responded quickly.

"The telephone line from the safety area was still operational. Your people and the hostages are ready to be hoisted. Nick said to tell you that he and Wally are coming up on top—with Jimmie—whatever that means. Also wanted you to know his last name means '*graveyard*' in Finnish. Says he didn't want to ruin your day when you first met. Lubo got out on the first chopper. He'll be all right and Jimmie's sister is going to make it."

Quinn pictured his two volunteers. Grandfather and grandson, coming up the way they went down—*niita* cradled between them— finally bonded in combat—and death . . . .

"GATEKEEPER to all hands. We're on our way! Have some of our men meet us at the top of the shaft."

Lashing safety cables around their waists, Quinn and Karin began the laborious climb up the narrow iron safety ladder of the main shaft. Ten minutes later, arms, legs and bodies aching, they were greeted by a group of smiling young Marines and their Norwegian counterparts that included blue-eyed young women carrying carbines. Sergeant Maki came out of the crowd and put his good arm around Karin.

The Head Frame was covered with shimmering hoarfrost. Quinn and Karin blinked in the bright sun and gazed upwards at the iron frame glistening in its white sheath.

Karin insisted that she earned her right to the final march. Quinn hesitated only momentarily and traded his M3 for an M16. Sergeant Maki, pride in his eyes, gave his to Karin.

They began the trek up the slope in the path left by Arbatov's command track. Their begrimed and bloody parkas snapped in the northern wind like spectral battle streamers. The Russian had chosen the most direct route to the high ground north and west of the Head Frame. Ahead of them, it crawled slowly toward the objective Quinn was sure Arbatov was seeking—line-of-sight access to *any* microwave tower.

The deep, authoritative rumble of Pratt and Whitney engines suddenly came out of the east. They both turned to see the low flying aircraft with the modified belly slide overhead. Seconds later the vintage bomber rotated up, engines snarling as ganged throttles hit the firewall. A massive plume of water cascaded onto the personnel carrier in a direct hit. A rainbow appeared and was gone in a flash. Before the first plane disappeared to the west, the second silver fire bomber followed, repeating the performance. Two minutes later a Bell Jet Ranger helicopter swooped in over the tree line. Hovering directly above the Command Track, its liquid cargo poured out of a 750-gallon neoprene bag in a well-aimed stream. The water froze on contact in the now Arctic air.

Arbatov's APC managed to find the high ground as the ice formed a shell over the entire unit.

Quinn asked Holmes to order in one more load from the chopper. Using borrowed Norwegian binoculars, he assessed his handiwork then handed the glasses to Karin. They moved on, closing the gap. Two hundred yards away, they saw the tracked vehicle grind to a halt on the high ground the Russian needed for access to the microwave tower and communications over a wide area of the U.S. and Canada—and if he made good his transmission—perhaps the world.

Water from the airborne tankers and helicopter had penetrated the carrier's tracks, freezing and jamming the driving mechanism. The APC took on the look of a mastodon frozen in Siberian ice.

The dish antenna, like the vehicle itself was now locked in ice following the second chopper dump. Quinn activated the portable radio. "KHE SAHN, Can we to talk to the Command Track?"

"They should be guarding one or more frequencies, Colonel. You might have to go through them all."

Just then, Quinn saw a wire laying on the crusted snow off to the right. The black wire led directly to the APC on the hill. Lifting it, he saw it was the landline, once attached to another unit, bayonet connector still in place. He screwed it into his handheld unit, realizing communications with Arbatov—if possible to make at all—would be between the two of them alone.

They pushed on, walking toward the sun, cold winds flapping the makeshift bed-sheets, another throwback to *Talvisota*. They paused as a final load of water splashed over the now totally encased carrier. To Quinn, it looked like a giant Japanese ice sculpture he'd seen in Sapporo, on Japan's northernmost island. He laughed to himself as they moved through the squeaking, crunching snow. *Polar shift, damn. It's ironic, using ice to stop armor.* He looked back over his shoulder. Three tanks lumbered down the main road, ice-covered but moving. He saw combat engineers already high on the Head Frame, cutting away bent steel with torches and re-threading the cable through the sheave to make the cage hook-up. Quinn and Karin stopped one hundred yards from the immobilized armor, taking cover behind a graywacke bolder as white camouflaged troops ringed the area below the hill.

Quinn glanced up, then pointed with his M16. Gossamer contrails created by dozens of fighters created a pattern of white lace against a delft blue sky.

He brought the transceiver up and tried the first channel. "Arbatov, this is Quinn. Do you read me?"

There was no answer.

He lifted his binoculars and scanned the ice-clad APC. He saw the thin white whip antenna almost hidden against the background of snow and ice and knew it must be linked to Arbatov's Transceiver. He motioned Karin forward, pointing toward the antenna. He spoke to

Ernie, advising him of their next move. Dropping to their knees, they looped the straps of the weapons around their arms in a hasty sling and took careful aim, firing quick bursts.

Ice and snow flew away from the base of the antenna. Checking again with the binoculars, Quinn looked at Karin smiling; "Up six and left six clicks Annie Oakley." They fired again—the sounds ricocheted in the air.   Quinn checked again with the binoculars. Only the stub of the antenna was visible. They had eliminated a major problem.

He looked at Karin with a smile. "I now know what *Huhta* means in Finnish . . . how about *Maki*?"

Karin wiped a blond wisp of hair from her forehead. "*Maki* means 'hill.' Grandma said our early forefathers lived on a hill to better defend their land . . . ."

Quinn looked around them. From their position they could see long distances. He laughed softly. Karin's eyes sparkled.

He helped her up from her kneeling position. "Makes sense Ms. Maki—a *Lotta* on the hill—fighting for her country . . . ."

Quinn moved forward, motioning Karin to stay behind the protective boulder.

Switching to the next frequency in order, Quinn tried again and then moved to the third frequency. "Colonel Arbatov!"

"You have Arbatov!"

Quinn detected a note of resignation in his adversary's voice.

"Arbatov, you are totally encased in ice. If you continue to run your engine, you'll die of asphyxiation. Turn it off and you'll freeze. Do you understand!"

"I understand very clearly my options Marine. My congratulations to a worthy opponent. You seem to have anticipated my every move."

"That may be Arbatov! I'm asking you to surrender—data intact. You can come out through the escape hatch on the floor of the APC— order your men to surrender."

"My men, Colonel Quinn, are all dead. If you and your men didn't kill them they died by their own hand."

Quinn shuddered at hearing the message.

A vicious gust of wind blew ice crystals in Quinn's eyes. He brushed the snow away with his gloved hand.

"You guessed correctly my last move Colonel Quinn. I can transmit to you on the tank inter-connect line but my primary transmitter indicates loss of antenna and—the hatches are frozen shut."

"Whom did you intend to transmit to Arbatov? What message did you want to send—to whom?"

Quinn was sure he knew Arbatov's mind by now and if not his motivation—his message and— audience . . . .

"If we couldn't pass the information from DANTE'S ORCHARD to our own people via the satellite you've apparently disabled, I was willing to share it with the *entire* world—something your country would most likely not do."

"I can't say, Colonel Arbatov. But I can guess that if you transmitted the ORCHARDS' message to the world, there could be great chaos—Pandora's Box to the hundredth power."

"Yes, a *'no-win proposition'* as you Americans call it. Should we have succeeded with our original plan and kept you and the major powers from the information—at least one small group of people could have survived to create a better world."

The setting sun silhouetted the command track standing out in stark relief against the red-orange sun.

"Colonel Arbatov, why the whip marks?"

"Start with the Fifth Tractus. Then add 4000 years of suffering."

Quinn shook his head at the enigmatic answer. "Your symbol—a cross with crescents within a star. What does it mean?"

"The Magen David, Colonel Quinn . . . ."

"The *magen*?"

"The Star of David—place it on the anvil of time and destiny and given the white heat of persecution, it can be hammered into many different shapes—a crescent moon–a cross . . . ."

Frustrated by the obtuse answers, Quinn persisted.

"You said something in Russian to the young woman while still in the mine." Speaking slowly, Quinn repeated Karin's words; 'You-moy-ya, shi-zin-ee—what does it mean? What does it signify?"

Quinn thought he heard the Russian laugh softly.

"What is your birth date Colonel Quinn?"

Puzzled, Quinn responded with the date.

"Ah, 'on the cusp' as people say. You and I Marine—are virtually the same age."

Frustrated, Quinn interjected, "What's that got to do with anything?"

"You speak Russian well. 'U moyey zhizni,' Colonel, means 'within my lifetime'

Staring at the frozen hulk and the sun dropping rapidly to the horizon, Quinn felt the icy cold—not wind driven but from foreboding.

The metallic voice of Arbatov continued: "Partial fruits of DANTE'S ORCHARD are now in one person's hands—yours! You know what the future holds in terms of time. You have a great responsibility. Some day, in one of Dante's realms, I hope to discuss with you how you handled such world shattering responsibility. U moyey zhizni—Colonel Caleb Quinn—can be bent to mean 'by your next birthday.' Arbatov paused, "The remaining fruits—the new world order—geographically—the LMRP—I take with me."

Quinn heard the hollow, ominous "click" and knew instinctively he'd heard the last of Arbatov—that the end was near for soldier, priest and penitent. In his mind, he could see the final absolutions taking place inside the APC.

Moving back behind the boulder, putting a protective arm around Karin, he pulled her down into the snow.

Changing channels, he spread word for Marines, Norwegians and Guardsmen ringing the site to take cover and aircraft above to clear the area. Seconds after his order went out, sparkling ice shards rose in the air. For one brief instant, the snow-covered knoll was brighter than the sun behind it. The massive concussion hit like a hammer blow. Chunks of white-hot steel ricocheted off the shielding boulder then sizzled in the snow behind them. The blast created a miniature blizzard as the explosion's echo rolled over algos oros, the 'mountain of pain,' like the mournful peeling of funeral bells. The fading tocsin reminding Quinn of the taps at Arlington—for his son Kyle. The words from a poem last read in college filtered through his mind—about the end of the world—words about fire and ice.

Echoes vanishing, he slowly rose to his feet lifting Karin to hers. They swept the snow from their blood and grease-stained parkas. Nothing remained of the ice-encased troop carrier or the fruits of

276

DANTE'S ORCHARD except a blackened spot on a shimmering field of white.

Quinn stared at the scorched rock that had only moments ago been a caisson crypt of iron and ice. He held Karin close to his side as the bitter cold regained dominance. Pale rose in color the sun was about to kiss the western horizon. Reaching into his pocket, he pulled out the convoluted symbol created from magen, moon and cross and felt a stabbing cold shudder throughout his body—not from the penetrating winds of the Alberta Clipper sweeping over the frozen landscape but from a deeper, more insidious terror.

"*U moyey zhizni*," he muttered under his breath, pulling Karin even closer.

A Fifth Force *was* assembling.

He alone on the planet earth knew *when* . . . .

The End

April 12, 2000